D0966313

Animal Experimentation

The Consensus Changes

ANIMAL EXPERIMENTATION

The Consensus Changes

Edited by
Gill Langley

First published 1989

Published in the USA, its Dependencies, and Canada by
CHAPMAN and HALL
The scientific and technical imprint of
Routledge, Chapman and Hall, Inc.
29 West 35th Street,
New York 10001–2291, USA

Typeset by Vine & Gorfin Ltd, Exmouth, Devon
Printed in Hong Kong

ISBN 0–412–02401–2 (hardback)
ISBN 0–412–02411–X (paperback)
Library of Congress Cataloging-in-Publication Data also available

Contents

619
a 598

Preface

This book addresses practical and philosophical issues regarding the use of animals in biomedical research, testing and teaching. It does so with the aim of presenting in some detail, from numerous perspectives and at different levels, facts and arguments to encourage scientists— whether researchers, doctors, veterinarians, laboratory animal technicians or science students—to reconsider their views about animal experiments. The intention is that fresh, thoughtful insight will be stimulated, generating a deeper concern for the many millions of animals involved worldwide.

The contributors to this book are recognised authorities in their fields, which range from animal-rights philosophy to the reduction of suffering in laboratory animals; from human–animal relations to the legislative scene; from science policy to humane research; and many areas in between. The starting points for the reader are contained in the first three essays. Mary Midgley digs at the roots of the human relationship with other animals: Do we consider ourselves part of the animal kingdom when it pleases us, and assume we are separate from it when expediency requires us to make a delineation? How does this schism affect our view of laboratory animals? Professor Tom Regan presents, logically and coherently, the rights argument as the basis for our dealings with animals, rather than a utilitarian view, and he uses scientific research as his illustration of the principles involved. Drs Margaret Rose and David Adams review, in some detail, the biochemical, physiological, anatomical and behavioural evidence that other animals are capable of feeling pain, and discuss the related question of suffering. Prompted in part by the requirements of new legislation to categorise the severity of procedures on animals, researchers and veterinarians worldwide are now concentrating on these issues.

The middle section of the book deals with the nuts and bolts of animal experimentation itself. Dr Erik Millstone, in his essay on science policy in the context of toxicology, has written a perceptive analysis of the bias and inconsistency which underlie many decision-making processes. Dr Robert Sharpe has marshalled extensive evidence for his case against reliance on animal experimentation as predictive for human medicine. Clive Hollands has documented examples showing that scientific research on animals can be, surprisingly often, little more than a 'trivial pursuit', and he expresses the view that it is the responsibility of all of us, scientists and lay-people alike, to condemn the trivial use of animals in laboratories. Dr Martin

Stephens brings us up to date with advances in humane research technology and its increasing potential to replace animal experiments, as well as discussing the factors that have encouraged and hindered progress in the field. Dr David Morton offers advice and information on ways in which experiments can be planned, conducted or modified to decrease the suffering of animals.

The last two essays look at responses by scientists and legislators to the changing consensus on animal experimentation. My contribution pulls together a number of threads which suggest that the scientific community has been reluctant, over many decades, to face the serious moral and practical issues raised by animal-based research. I attempt an analysis of these problems, and suggest that a more sensitive and open-minded approach is overdue. Finally, Dr Judith Hampson reviews the recent changes in British, American and European animal experimentation laws, and assesses the degree to which different legislative systems might be successful in achieving their stated aims.

It has not been an easy task to assemble, in one volume, essays from authors whose personal positions on the subject are widely disparate. Professionals who are known for a particular stance at some point on the spectrum of views on animal experimentation are understandably nervous of seeming to compromise their position. For example, a researcher whose essay might appear within the same book covers as that of an animal-rightist risks being criticised by colleagues for 'going soft' or, worse, for giving credibility to the 'extremists'. Anti-vivisection contributors found in the company of scientists face the accusation that they are compromising their ideals or are in the pay of the drug companies. As the old Chinese proverb so rightly said, 'He who walks in the middle of the road is hit by chariots in both directions'.

These reactions are predictable because the issue of animal research has moved from the wings into the centre-stage spotlight of political, social, scientific and moral concern. Sides have been drawn up, emotions have run high, reputations and livelihoods have sometimes been at stake. That the contributors to this book have agreed to rub shoulders with each other is not so much a result of editorial tact and persuasion on my part (although some of this was required!), as of a realisation that the problem of animal experimentation now, more than ever before, demands an overview, clear thinking, analysis and constructive action by all concerned people, no matter what their shade of opinion. One thing which all the authors in this volume share is a desire to see the suffering of laboratory animals, mental and physical, ameliorated. It should be clear from the foregoing, but I must state it anyway, that the arrival of this volume on the bookshop shelves does not necessarily signify that its contributors agree with each others' detailed views on the subject.

I must also make clear that my own position is not a neutral one. I am, and have been for some 11 years, a committed proponent of animal rights and anti-vivisection. During the preparation of this book it has been difficult, quite often, to withhold the editor's red pen where authors have made statements which clash directly with my own views. However, I believe that to convey an important message successfully requires the presentation of convincing arguments at every and any level where they might be effective. Among readers of this volume, those whose livelihood depends on animal experimentation are unlikely to become supporters of animal rights overnight, no matter how persuasive the argument (which must nevertheless be presented). But such researchers will, at least, be open to advice on modifying their experimental protocol to reduce animal suffering. Scientists whose daily work experience demands an emphasis on objectivity may be enticed into letting a little more feeling and empathy into their relationships with laboratory animals. Those who are already committed to an animal-*welfare* position may see the logic of the animal-*rights* philosophy. Science students, on reading this book, may become aware of the desensitisation which occurs in science education, and see the wisdom of resisting it. Veterinarians involved in animal research might feel moved to review the meaning of the veterinary 'Hippocratic oath', which binds them to consider as a priority the interests and well-being of animals. Concerned lay-people will find here an exposition of the contemporary debate on animal experimentation, with the use of technical terms only where this is unavoidable.

The literature on animal experimentation generally divides into clear categories. The obviously pro- or anti-vivisection books, such as Paton's (1984) *Man and Mouse* or Ryder's (1975) *Victims of Science*, may be read by followers of the 'other camp', but usually only with the intention of finding flaws in their opponents' argument. The detailed philosophical approach, such as Regan's (1984) excellent *The Case for Animal Rights*, can nevertheless be daunting for non-philosophers. Several books have tackled a range of animal-orientated issues, including factory farming and bloodsports, from a strongly pro-animal viewpoint, such as Singer's (1985) *In Defence of Animals*, but these may have a limited appeal for scientists. Rowan's (1984) *Of Mice, Models and Men* and Sperlinger's (1981) *Animals in Research* contained much of interest for pro- and anti-vivisectionists. This book follows and expands on those earlier works, in addressing a primarily, but not solely, scientific readership on the topic of animal experimentation; in explaining the limitations of animal models in medicine and toxicology; in presenting a wide spectrum of expert views on animal welfare and rights, in a factual and non-combative way; and in bringing readers up to date with practice and philosophy, policy and politics. We also hope to change hearts and minds, and encourage a new and just perception of animals as individuals whose interests and needs deserve our careful attention. The task is probably impossible, but the arguments for concern and compassion which are marshalled in this book are, no matter what your own present position, compelling.

Hitchin, Herts, 1989   Gill Langley

REFERENCES

Paton, W. (1984). *Man and Mouse: Animals in Medical Research*, Oxford University Press, Oxford

Regan, T. (1984). *The Case for Animal Rights*, Routledge & Kegan Paul, London

Rowan, A. N. (1984). *Of Mice, Models and Men: A Critical Evaluation of Animal Research*, State University of New York Press, New York

Ryder, R. D. (1975). *Victims of Science*, Davis-Poynter, London; revised and reprinted (1983), National Anti-Vivisection Society, London

Singer, P. (ed.) (1985). *In Defence of Animals*, Basil Blackwell, Oxford

Sperlinger, D. (ed.) (1981). *Animals in Research: New Perspectives in Animal Experimentation*, John Wiley, Chichester

The Contributors

Dr David Adams has worked in general veterinary practice and has studied immunology at Oxford University. His present position is at the CSIRO Pastoral Research Laboratory at Armidale, New South Wales, Australia, where his major interests include immunology, animal husbandry and animal welfare.

Dr Judith Hampson has a background in biology, and for six years worked at the RSPCA. She is now an independent consultant in laboratory animal welfare, her special interests being legislative reform, education and the philosophy of animal rights. She is a member of the UK Government's Animal Procedures Committee, and an adviser to the EC Commission on the implementation of the EEC Directive on the use of animals in research.

Clive Hollands has worked for human as well as animal welfare for many years. He is Consultant to the Scottish Society for the Prevention of Vivisection, Secretary to the St Andrew Animal Fund and also to the Committee for the Reform of Animal Experimentation. He is a member of the Home Secretary's Animal Procedures Committee and an Honorary Associate of the British Veterinary Association. His views on joint action and compromise have made him a controversial figure in the animal-welfare movement.

Dr Gill Langley has a doctorate in neurochemistry from Cambridge University and experience of tissue-culture research. An animal-rightist and a vegan since 1978, she is scientific adviser to the Dr Hadwen Trust for Humane Research and to Animal Aid. She is also a freelance writer and consultant on animal experimentation, humane research, alternative medicine, and diet and health.

Mary Midgley is a philosopher with a special interest in evolution, in the relation between our species and those around it, and in the way in which our beliefs about these things can affect our practical conduct. About the treatment of animals, she particularly welcomes the present tendency for co-operative discussion to replace barren debates between extremists, and believes that this clearer and less combative thinking may lead to necessary and useful changes in practice.

Dr Erik Millstone is a lecturer at Sussex University's Science Policy Research Unit. In the course of his work on the testing and control of food additives, he has explored some aspects of the scope and limits of toxicology and the value of animal tests for chemical safety.

Dr David Morton is a research scientist and a laboratory animal veterinarian, with a particular interest in animal welfare and the assessment of pain and distress in laboratory animals.

Professor Tom Regan is a Professor of Philosophy at the North Carolina State University, USA, and president of the Culture and Animals Foundation. He is author and editor of over 20 books, many of which address the issue of animal rights. His *Case for Animal Rights* is the definitive work on the subject.

Dr Margaret Rose is a veterinarian with a PhD and has been involved in biomedical research for 20 years. She is a Senior Project Scientist at the University of New South Wales, Australia, and has been involved in animal-welfare issues since 1975. She serves on a number of committees, including the Animal Welfare Advisory Council and the Australian Council on the Care of Animals in Research and Teaching, and is Chair of the New South Wales Animal Research Review Panel.

Dr Robert Sharpe is a scientific adviser and consultant to the British and international animal-rights movement. He was formerly a research chemist at the Royal Postgraduate Medical School in London and also served as Director of the Lord Dowding Fund for Humane Research, an organisation which promotes and sponsors the development of alternatives to animal experiments.

Dr Martin Stephens is Director of the Laboratory Animals Department at the Humane Society of the United States. Prior to joining the HSUS staff in 1985, he was a consultant to several animal protection organisations and wrote critiques of animal experimentation in particular fields of research. His own research and training is in zoology and he received his PhD in biology from the University of Chicago in 1984.

1. Are You an Animal?*

Mary Midgley

'He who understands baboon would do more towards metaphysics than Locke.'

Charles Darwin's notebooks

THE QUESTION

If anthropologists from a strange planet came here to study our intellectual habits and customs, they might notice something rather strange about the way in which we classify the living things on our planet. They would find us using a single word—animal—to describe a very wide range of creatures, including ourselves, from blue whales to tiny micro-organisms that are quite hard to distinguish from plants. On the other hand, they would note also that the commonest use of this word *animal* is one in which we use it to contrast all these other organisms with our own single species, speaking constantly of 'animals' as distinct from humans. It might strike them that in virtually every respect gorillas are much more like ourselves than they are like (say) skin parasites, or even worms and molluscs, so that this use of the word is rather obscure.

These two distinct ways of thinking are my topic here. Both are used readily in everyday life. If a small child asks us what an animal is, we are likely to choose the first meaning, and our answer will probably be wide, untroubled and hospitable, more especially so if we are scientifically oriented people. We shall explain that the word can include you and me and the dog and the birds outside, the flies and worms in the garden and the whales and elephants and the polar bears and Blake's tiger. In other contexts, however, we may find ourselves using the word very differently, drawing a hard, dramatic black line across this continuum. 'You have behaved like animals!' says the judge to defendants found guilty of highly complicated human social offences, such as driving a stolen car while under the influence of drink. What is

*Some parts of this essay have grown out of a paper read at the World Archaeological Congress in September 1986, which will be published in its records under the title 'Beasts, brutes and monsters'. I am grateful to the organisers of that Congress for permission to reprint these passages.

1

the judge doing here? He is, it seems, excluding the offenders from the moral community. His meaning, as widely understood, is something like this: 'You have offended against deep standards and ideals which are not mere local rules of convenience. You have crashed through the barriers of culture, barriers which alone preserve us from a sea of abominable motivations. The horror of your act does not lie only in the harm you have done to your victims, but also, more deeply, in the degradation into which you have plunged yourselves, a degradation which may infect us all.'

I think this is a fair interpretation of such common remarks—an interpretation which covers their main points, though of course in such an emotive and disturbing matter more is certain to be involved, and this notion of an 'animal' clearly carries us into somewhat mysterious areas of our ill-understood habits of symbolism. This is not some casual ambiguity, which could be remedied by updating the dictionary. By the very nature of the case it touches on matters which it will frighten us to think about. In its second use—the one which excludes humanity—the word 'animal' stands for the unhuman, the anti-human. It symbolises the forces that we fear in our own nature, forces that we are unwilling to regard as a true part of it. By treating those forces as non-human, it connects them with others that we fear in the world around us—with floods, earthquakes and volcanoes—and it thus dramatises their power, but it also enables us to disown them. It implies that they are alien to us, they are outside us and therefore incomprehensible. We insist that we are not responsible for these motives. But the peculiar kind of horror that they produce suggests that there is a lot of bad faith in this insistence, that we are not altogether convinced of their externality. We see these frightening forces not just as outside dangers like earthquakes, but also as dangers within us, seeds that lie hidden in our own nature, and that may at any time develop if outside offenders are allowed to encourage them by their example.

That, I am suggesting, is the traditional attitude, both in our own culture and to some extent in many others, to what an 'animal' is. The second part of it is not indeed often spelt out these days, but then it does not need to be because it is a powerful, ancient, imaginative background that works by being taken for granted. Now clearly, any concept riven by an ambivalence as deep as this is not going to yield us a single, clear meaning, but a thicket of instructive confusions. The word 'animal', though used as a perfectly good term of science, does most of its work in areas which are not in the least detached or scientific, and this makes it a very illuminating example of the way in which our scientific and our everyday thinking interrelate. Its two usages play, I believe, a very important part in that thinking, notably in forming our communal self-image—our notion of the kind of beings that we ourselves are. Whatever propaganda humans in a particular age want to put about concerning themselves demands, and gets, corresponding alterations in the typical notions entertained of non-human animals.

OUTER DARKNESS

These conflicting ideas about the meaning of the human–animal frontier are very ancient, but their clash is sharpened up today by the notion we now hold of ourselves as thoroughly scientific beings— individuals too clear-headed and well-organised to use blurred or ambivalent concepts. The concepts we need to use for everyday life are, however, often in some ways necessarily blurred and ambivalent, because life itself is too complex for simple descriptions. For instance, the concept of a friend is not a simple one, and people who insist on oversimplifying it cannot keep their friends, nor indeed be friends themselves, because they do not properly understand what a friend is. The same difficulty constantly arises about many concepts in biology— for instance evolution, adaptation, organism and indeed life itself. Oversimple definitions of terms like these have again and again distorted science. The standards of clarity that we manage to impose in our well-lit scientific workplaces are designed to suit the preselected problems that we take in there with us, not the larger tangles from which those problems were abstracted.

I am suggesting that the ambivalence just noticed in our attitude to the species barrier is one of these large tangles, and needs to be taken seriously. Our twofold use of words like 'beast' or 'animal' is not just a chance ambiguity that we can set right by policing usage. We cannot, for instance, merely rule that only the first meaning (the inclusive one) is scientific, on the grounds that it accords with current taxonomy and the theory of evolution, and therefore that it alone has a right to survive. For one thing, objectors might at once say that the second meaning is really the scientific one, because non-human animals must be used as subjects for scientific experiments, and must therefore be classed separately from people in order to make it clear that they have no rights that would prevent this. Here 'scientific' would mean merely something like 'necessary for the practice of science'. More subtly, however, others would certainly say that, in the interests of truth itself, the scientific approach demands that the difference between humans and all other animals should be treated as paramount, because these creatures are in fact beings of a distinct kind, much more like machines than they are like humans.

This approach arose in the first place out of Descartes' simple-minded mechanistic belief that animals are actually unconscious, and it was immensely encouraged by crude behaviourist psychology. It has been radically undermined in this century by the advance of ethology, which has supplied for the first time solid, incontrovertible evidence that the lives of many other social animals resemble human life far more closely than had ever before been believed, and cannot be properly described without using many concepts suited for describing the behaviour of humans. No serious and well-qualified reader can dismiss the accounts given by Jane Goodall, Arthur Schaller, Dian Fossey and their many colleagues as merely sentimental or 'anthropomorphic' wish-fulfilment. Plainly, these people are scientists. On the other hand, the speculative excesses of early, metaphysical Behaviourism under Watson have not worn well, and its central doctrines do not now seem 'scientific' at all. More generally, too, mechanism itself has for some time been losing ground since machine models have proved less and less useful in physics, which was the field where they originally gained their prestige. Physicists are now often rather surprised to find that biologists still show so much respect for these machine models, with which they themselves have become somewhat disillusioned. But these changes are slow. Undoubtedly, the set of ideas that ruled at the beginning of this century still has great influence. Many people do still habitually think that mechanistic explanations are indeed always more 'scientific' than ones which use concepts appropriate to a human context, even in situations where they demonstrably fail to do any effective explaining.

It would not, then, be easy to arbitrate between the two uses of the word 'animal' merely by deciding which was the more scientific. But even if that decision could be made, usage could not be forced to conform with it, because people are in any case not always talking science. Both usages are common because both are emotive. To think of ourselves seriously as animals is to regard the other animals as our kin; it inevitably leads us in some degree to welcome them, to identify with them, to see their cause as our own. That, indeed, is just what people find both attractive and frightening about this way of thinking. We have to notice that, in general, value concepts are not actually tidily separated from factual or descriptive ones, however much it might simplify our arguments if they were. There are nearly always conceptual links, and indeed this question about the species barrier is a good example of such an irremovable connection. How we regard this barrier cannot be a neutral matter for us. To some extent and in some ways, the idea of it is bound both to suggest to us that we belong in a wider sphere, and also to indicate for us the frontier of value, so that 'human' becomes an important term of praise.

This last thought is as unavoidable as the other. Since human beings have to live a human social life, which they often find hard, especially in childhood, the notion of the great, dark, non-human area outside is bound to strike us in some ways, right from our earliest days, as something forbidden, alien and probably frightening. This area includes, in uncertain relation, the unacceptable parts of our own nature and the entire natures of the other animals around us. That is why an obvious and familiar kind of horror attends situations in which human beings are treated, as we say 'like animals'—for instance, where they are herded into cattle-trucks or left to starve or, most particularly, are eaten without scruple. Similar horror is conveyed by the thought of their 'behaving like animals', and this, as we have noticed, may simply mean 'not how human beings are supposed to behave in our culture', with a special emphasis on the kind of motives involved. The idea of a mixed, partly human monster such as the Minotaur symbolises this special kind of fear and disgust.

I have begun by stressing the hostile, exclusive half of our divided attitude, because I think we often do not realise quite how much it influences us. The sense of drama that attends controversy about human origins, and the way in which new speculations about the source of 'human uniqueness' spring up full-grown on the heels of even the slightest archaeological discovery, show clearly how nervous people still are about the idea of a 'missing link' that might bestride the species barrier. People are afraid not just of finding that they have discreditable ancestors, but of something that those ancestors might reveal about human nature today. We know we do not fully understand our own nature. Of course, we have certain working notions about it, but continually we find difficult cases cropping up in which these notions fail us, precipitating us into theoretical and (still more obviously) into practical disaster. If we think seriously about it today, we are surely likely to find ourselves still in agreement with the view of humanity that Alexander Pope expressed in the opening lines of his *Essay on Man*:

Perched on the isthmus of a middle state,
A being darkly wise and rudely great . . .
He hangs between, in doubt to act or rest,
In doubt to deem himself a god or beast, . . .
Sole judge of truth, in endless error hurled,
The glory, jest and riddle of the world.

This is a disturbing picture. During the Enlightenment, therefore, thinkers made great efforts to simplify and domesticate it by treating the darker, more mysterious, aspects of human life as mere historical accidents, effects of unnecessary moral and political failures, 'artefacts of the system'. If they had succeeded in establishing this view—that is, if they had managed to abolish these blots by altering the educational system—then perhaps we might today be able to look at other animals more dispassionately, as beings quite separate from ourselves, which we were not called upon to judge as either good or bad. But, in spite of many important minor gains, of course they could not produce that total revolution. Human conduct did not dramatically improve, nor were the dark places of the soul found to vanish. At the same time, however, the advance of science connected human beings more firmly than ever with the other animals through the theory of evolution.

Darwin himself responded positively to this change. It seemed to him obvious that the new ideas implied a strong and significant continuity between human nature and the nature of other creatures. A scientific approach therefore now called for the end of all prejudice against a serious, dispassionate comparison between their psychologies. The best prospect for understanding human motivation lay in assimilating the conceptual schemes used for these two studies, and in developing both through the systematic comparison between them. For this purpose, Darwin was prepared to raid the full range of psychological concepts that have been developed for describing human feeling and behaviour—a range so rich and varied that, if intelligently handled, it can be expected to provide suitable ways of describing the traits that we share with other animals, as well as those that are peculiar to ourselves. Darwin used this method effectively himself in *The Expression of Emotions in Man and Animals* (Darwin 1872), and it was later taken up and developed for the founding of modern ethology by Niko Tinbergen and Konrad Lorenz.

After Darwin's death, however, the tide turned against all such thinking. Behaviourist psychology did indeed officially treat humans and other animals as similar, but it did it in exactly the opposite way from Darwin, by treating both as insensate machines. Early, dogmatic, metaphysical behaviourism ruled that everything equally was a mere object; there was no such thing as a thinking subject, and the whole idea of 'consciousness' was merely a superstition. The fearful confusions that resulted from this idea led John Watson's heirs to abandon it, but unluckily they did not openly think through and set straight the wider metaphysical notions that had first produced it, but merely cursed metaphysics and withdrew to slightly safer ground. For human psychology, they continued to use mechanical models alongside nominal admissions that subjectivity was present, without any real attempt to resolve the clashes that these two discrepant ways of thinking constantly produced. But for animal psychology, pure mechanism still largely ruled, because it was still held up as 'scientific'. (For a full and balanced account of this fascinating story, see Boakes (1984).) Sociologists and anthropologists, however, continued mean-while to treat human beings as unique, usually denying strongly that comparisons from any other species could possibly be relevant to them.

Thus from the time of Darwin's death until the development of ethology in the present century, most of the scholars whose studies neighboured the species barrier viewed the gap as unbridgeably wide, and the behaviourists who thought otherwise did so because they assimilated both parties to machines. Moreover, behaviourists and sociologists alike largely denied the presence of inborn behavioural tendencies in humans. Many things contributed to produce this change from Darwin's position, but among the intellectual factors involved, probably the foremost was the increasing specialisation which went with the professionalisation of science. Social and physical scientists increasingly treated each other as alien tribes, and were not surprised to find that they were thinking on different lines.

I do not think that we can fully understand this change without noticing also wider social and emotional factors as well as professional ones. The notion of an animal is, as I have suggested, a deeply and incurably emotional one, about which we cannot be personally neutral. If we do not respond to it with a positive sense of kinship, as Darwin did, we are almost certain to do so with the hygienic, rejective horror already mentioned. Darwin was exceptional, not just in his scientific ability, but in his awareness of the symbolic forces that cluster round such topics, and in the bold and generous spirit which often enabled him to make good choices among them. Once his approach was written off as amateurish, scientists who supposed themselves to be thoroughly detached and impartial often responded very confusedly to these symbolic cues. That, I think, is why chronic, endemic exaggeration of the differences between our own species and others became for a time widespread. (I have discussed its distorting effects in Midgley (1978, 1981a, 1984) and elsewhere.) This exaggeration was especially dogmatic in the social sciences, but biologists too seem often to have accepted it without much question as part of a scientific attitude, and have been willing to agree that reasonings belonging to their own discipline could not possibly apply to the human race—until the ethologists began to suggest otherwise.

WIDENING THE GAP

This problem tended to be seen as one of *parsimony*—that is, of how to avoid adding anything to the notion of an animal as simply a machine. Given that initial starting point, the addition of 'consciousness' was viewed as a piece of extravagance, and any further attribution of subjective attitudes such as purpose or emotion appeared more extravagant still. Behaviourism had originally taken the same supposedly austere line about human beings, and in principle it continued to do so, but this method worked out so badly over most of the field of social science that it never became supreme there. It was fairly quickly realised that the machine model is just one possible way of thinking, with no special authority to prevail where it does not give useful results. On the non-human scene, however, mechanism was not seriously questioned because scientists had not yet seen its general disadvantages, nor had they paid sufficient attention to animal behaviour to see that it worked just as badly there.

Thus there was a remarkable discrepancy between what was treated as a parsimonious explanation for a piece of human behaviour, and what could count as such when the behaviour was that of some other animal. The practice was that, in the human case, the normal, indeed practically the only, licensed form of explanation was in terms either of culture or of free, deliberate choice, or both. Anyone who suggested that an inborn tendency might be even a contributing factor in human choices tended to be denounced as a fascist. The burden of proof was accordingly laid entirely on this suggestion, and was made impossibly heavy. To put it another way, any explanation which invoked culture, however vague, abstract, far-fetched, infertile and implausible, tended to be readily accepted, while any explanation in terms of innate tendencies, however careful, rigorous, well-documented, limited and specific, tended to be ignored. In animal psychology, however, the opposite situation reigned. Here, what was taboo was the range of concepts that describes the conscious, cognitive side of experience. The preferred, safe kind of explanation here derived from ideas of innate programming and mechanical conditioning. If anything cognitive was mentioned, standards of rigour at once soared into a stratosphere where few arguments could hope to follow.

The tide in both areas has certainly turned, and I do not think the tradition will last anybody much longer. Nicholas Humphrey (1976, 1978) has floated the entirely convincing suggestion that consciousness and intelligence in social creatures must have evolved largely to deal with social problems rather than merely practical ones—a suggestion which makes the continuity with human life so glaring that ignoring it any longer is scarcely possible.

On the issue of parsimony which I have just mentioned, Donald Griffin (1984) effectively shifted the burden of proof, pointing out how odd it was to suppose it *more* parsimonious to account for highly complex and flexible behaviour by positing a program so elaborate that it can provide for every contingency, than to make the much more economical assumption that the creature had enough brain to have some idea of what it was doing. As he points out, the attempt to make pre-programming account for everything has only been made to look plausible by constant misdescription—by abstract, highly simplified accounts of what creatures actually do, accounts which have repeatedly been shown up as inadequate when observers take the trouble to record more carefully what happens.

Highly complicated performances by relatively simple animals can indeed be accounted for to some extent by positing that they possess inborn 'neural templates' which they use as patterns. But considering the skill and versatility with which they adapt these patterns to suit varying conditions and materials, it makes little sense to suggest that the templates reign alone and can, so to speak, work themselves:

> . . . Explaining instinctive behaviour in terms of conscious efforts to match neural templates may be *more parsimonious* than postulating a complete set of specifications for motor actions that will produce the characteristic structure under all probable conditions. Conscious efforts to match a template may be more economical and efficient. . . . *It is always dangerous for biologists to assume that only one of two or more types of explanation must apply universally.* (Griffin 1984, p. 116; emphases mine)

He cites the well-known case of birds which lead predators away from their broods by distraction behaviour, acting as if they could not fly properly until they have moved the threat well away from the nest, and then flying back in a normal manner. Scientists have gone to great lengths to account for this well-established practice without invoking conscious intention, by positing conflicts of inborn drives such as fear and parental concern. These conflicts are supposed to produce hesitant and contradictory behaviour, which then happens, by an incredibly lucky chance, to be regularly misinterpreted by predators as inability to fly. Griffin comments:

> The thoughts I am ascribing to the birds under these conditions are quite simple ones, but *it is often taken for granted that purely mechanical, reflex-like behaviour would be a more parsimonious explanation* than even crude subjective feelings or conscious thoughts. But to account for predator-distraction by plovers, we must dream up *complex tortuous chains of mechanical reflexes.* Simple thoughts could guide a great deal of appropriate behaviour without nearly such complex mental gymnastics on the part of the ethologist or the animal. (Griffin 1984, p. 94; emphases mine)

In this case the traditional explanation is particularly feeble, because plainly parents in very many species must actually engage in conflict behaviour on these occasions—but only with these particular species of birds does it take this form and so strangely mislead the predators. For these species, however, the mistake is regularly made by a wide variety of predators, although it is the business of all predators to understand well the typical behaviour patterns of their proposed prey. Moreover, the point at which the conflict behaviour unaccountably ceases and the bird flies home just happens to be one where the predator has been led far enough off from the nest not to go back . . . . This is surely an explanation that no-one would put forward except to save a dogma that is no longer worth saving. The dogma is that non-human animals cannot plan, and in particular cannot deceive. But there is by now plenty of evidence that they sometimes can, and there is no need for fantastic solutions of this kind to be devised for such problems.

WHICH COSTS MORE?

The question Griffin raises here is central. Why is it supposed to be more economical to account for the behaviour of animals *without* treating them as conscious? Why is consciousness regarded with suspicion as a sinister extra entity, instead of as the normal function of a developed nervous system? How could it be economical to *remove* such an obvious function from the brain when that brain already exists? What—more generally—does scientific parsimony usually require of us? Parsimony plainly does not have the purely negative aim of just leaving things out, of making explanations as simple as possible, for if it did the best explanation would always be the shortest. On these principles, the Biblical account of creation would excel all others, since it names only a single cause—God—and abstains from complicating matters by adding any details about his modes of working. We are, of course, sometimes forced to accept accounts as simple and general as this, where our ignorance is very deep, but such honest admissions of ignorance are not explanations.

Neither—again—can parsimony mean just refusing to use more than one pattern of explanation, economising on our basic methods of thought. That was indeed the idea which led the early dogmatic behaviourists to exclude all reference to subjective motives from their accounts of both animal and human behaviour. Their approach has been found unsatisfactory for human cases, because it involves ignoring a mass of relevant and useful evidence, and indeed it has proved scarcely possible even to describe the 'objective' evidence about human beings on its own, without constantly referring to the subjective aspect which forms an inseparable part of it. Behaviourist psychological methods did, however, impress many people for a time as 'scientific' because they used terms which were familiar in the physical sciences, and avoided ways of thinking unique to human psychology. Griffin rightly calls attention to the misleading effect of this deliberate imitation of another science, and the dangerous false reassurance that can be derived from thinking that this mere surface imitation makes one's methods scientific, when in fact one's distinctive subject matter demands a method of its own. The mere negative effect of removing subjective elements from a given explanation has no special value. What parsimony calls for is that we remove *irrelevant* elements. And it is not clear why subjective elements should be supposed to be irrelevant to behaviour.

Why is it believed that concern with subjective states is unscientific? One reason which seems to convince some people is a fairly simple confusion about the status of subjectivity itself, an impression that to study subjective phenomena is the same thing as 'being subjective'— that is, being tossed about by one's own moods and feelings. This seems to be the same mistake as supposing that the study of folly must be a foolish study, or the study of evil conduct an evil one, or in general (as Dr Johnson put it) that 'who drives fat oxen should himself be fat'. Behind this simple error there lies the rather more solid point that there is a difficulty in seeing how we can know anything about the subjective states of others. It is true and important that our knowledge of these states is limited. But if we really had no such knowledge our world would be totally different from what it is, and we should not possess any concepts for describing or understanding our own subjective states either. If we say that we never know at all whether anybody else is angry or afraid, or in pain, or aware of something, or expecting something of us, our actions will immediately give us the lie, and we know very well that to pretend to suspend judgement on such matters would in fact be mere humbug. If, for instance, a torturer were to excuse his activities by claiming not to know that his victims suffered pain, he would not convince any human audience. And an audience of scientists need not aim at providing any exception to this rule.

DILEMMAS ABOUT PRIMATES

If we accept Griffin's contentions as at least evening up the score on the issue of parsimony, are there any other considerations that ought to convince us that animals do *not*, in fact, think and feel as their conduct and the size of their brains makes it natural to suppose that they do? Or that their thoughts and feelings in particular situations are *not* roughly of the kind that we would expect them to be, when our expectations are based on human experience gathered over the ages, experience both of our own species and of those around it? Is there, for instance, any good reason to suppose that a baby rhesus monkey, when removed from its mother at birth and placed in a stainless-steel well, does *not* feel something like the same kind of misery and fear that a human baby might be expected to feel in the same situation? It is interesting to notice that language does not really seem to make much difference here. Most of us would not doubt that a human baby would feel these things, even though it could never tell us so. And in general, in dealing with babies, we never let their speechlessness make us doubt that they do have thoughts and feelings, because it is only possible to deal with them successfully if we do treat them as conscious. Babies, as much as human adults, insist on being treated as people, not as things; behaviourist scepticism that required a different method could lead only to disaster.

The same thing is true of baby apes and monkeys, and those who deal with them have constantly to act accordingly. This case is interesting, too, because of a dilemma that arises out of the justifications that have been given for such experiments. These justifications have centred on the claim that they threw light on the origins and nature of depression and other mental troubles in human beings. States such as depression are, however, ones in which subjective elements are of the first importance, and this is normally assumed to be true also of the history that leads to it. If the rhesus infants were really to be regarded as mere robots, crying only in a mechanical manner like unoiled machinery creaking (as Descartes' followers put it; see Rupke (1987)), could any useful parallel be drawn between their reactions and those of a human being? Even if they have sensibilities, but ones much simpler and less intense than those of humans, can the parallel be of any value? Because of the obscurities surrounding this point, it is not surprising that the long series of experiments of this kind seems in fact to have had virtually no consequences of value for the treatment of human mental illness (see Stephens 1986). The ill-effects of maternal deprivation were known before it started, and the further damage done by environments such as steel wells have little relevance since these things do not happen to humans. In recent years, increasing numbers of scientists have begun to be worried by this disturbing dilemma about primates, and to reason that if they are sufficiently like us to be really comparable, they may be too like us to be used freely as experimental subjects.

Ought we, then, to promote all primates—or at least the great apes— to the position of honorary humans, crediting them with human-like subjective states and according them human-like rights, while leaving the rest of the animal kingdom still outside in the darkness? This has been proposed, but a moment's thought shows that it cannot be the answer. There is too much continuity between primates and the rest. No single sacred mark picks the primate order out from all the others, as the possession of an immortal soul has been held to pick out the human race. If we think that rhesus monkeys are capable of having thoughts and feelings that deserve our consideration, then we must think the same of other mammals and birds and quite likely very many other creatures, too. Though the nature of their subjective states will doubtless vary vastly and often be obscure to us, their mere existence puts us in a relation with these creatures which cannot be the same as our relation with a stone or a tin tray. How close then are they to us?

DIVIDED FEELINGS

I have devoted much of this essay to discussing our ambivalence about this question because I think it is a very important factor, though a negative one, affecting all the positive conceptions that we form of other species. Insofar as it obstructs our free thinking on these subjects, it is something of which we need to be aware. How far it actually does obstruct is very much a matter of opinion, and the influence certainly varies a lot in different areas. Jane Goodall single-handedly has, I think, succeeded in transforming our view of the great apes, simply by showing what a high degree of scientific rigour can be combined with an entirely personal approach to the individuals studied, and how much the personal approach then helps the rigour in furthering our understanding, not just of these particular apes, but of animal and human nature altogether. She is, however, part of a much wider ethological tradition which has been working in this way across the board, and has profoundly altered our attitudes. About the primates in particular, this new approach has, as I mentioned earlier, begun to raise doubts affecting the ethics of experimentation, and is already beginning to change scientific practice. This change, however, has not yet got very far, and the most striking thing about the present situation is its extraordinary unevenness.

Quite often we are moved by a strong Darwinian or Franciscan sense of kinship with other creatures, which can be just as influential as the distancing and revulsion which replace it at other times. What is really worrying at present is the impression many people have that the revulsion is somehow 'more scientific' than the affection and respect. This idea rests on two very strange suppositions—first, that science ought not to be inspired by any emotion, and secondly, that disgust and contempt are not emotions, whereas love and admiration are. It would seem to follow that all enquirers who have worked out of pure admiration for their subject matter, from the Greek astronomers gazing at the stars to field naturalists who love their birds and beetles, would be anti-scientific, and ought if possible to be replaced by others who are indifferent to them, or who actively dislike them.

This is an attitude that nobody is likely to endorse once it is openly spelt out. In general, most people now admit that it is wrong to ill-treat animals unnecessarily. But reformers who want to draw attention to ways in which we seem to be ill-treating them have to use our existing moral language, which is of course largely adapted to describing relations between humans. When, therefore, it is suggested that we ought to be concerned also about the suffering of other animals, this idea can have the disturbing effect that I mentioned earlier—it can sound monstrous. This happens particularly easily when the creature in question is not one that is integrated into human life as a familiar pet or servant. People hearing protests on behalf of such creatures often take refuge from their scandalised reaction in laughter: 'Are you really making all this fuss about guinea-pigs—or pigs—or (still stranger) rats?'

SELECTIVE DESENSITISATION

All these cases have some features of interest. Rats are in fact lively, intelligent and sociable creatures, an opportunist species that naturally explores its environment, and is therefore capable of being bored when that exploration is frustrated. They are also able to respond well to human beings, as those who keep them as pets know. But their public image has of course been largely formed by their long history not as pets but as pests—that is, first as sharp competitors with us, from the dawn of agriculture, for access to stores of grain, and then as carriers of disease. Their tactless failure to grow fur on their tails also gets them a bad name by reminding many people of snakes, which are another symbolic focus for fear and hatred.

All this has made it easy for modern people to see rats as some kind of undeserving monster, and the projected fear that goes into describing a bad human being as 'a rat' serves to dramatise this notion yet further. Mice, being smaller, convey a slightly less vicious impression, but do not do much better out of it because what is smaller seems less considerable anyway. Moreover, a strange new twist was given to the rodent image in the heyday of behaviourist psychology, when rats and mice were so extensively used as standard experimental subjects that one researcher actually dedicated his book 'to Rattus norvegicus, without whose help it could never have been written'. This mass of 'rattomorphic' psychological theory supposedly applicable to humans is not now thought to have been vastly useful, but it did manage to do one thing. It fixed the notion of the rat itself as simply a standard object, a piece of laboratory equipment with the function of being used to test hypotheses, a kind of purpose-made flesh-and-blood robot. And it served to condition scientists to this view of the animal. This conditioning is partly visual, because anyone who frequently sees a stack of standard small metal cages, each containing one bored white rodent which is never seen otherwise occupied, will be liable to absorb this impression. It is, however, also verbal. In scientific articles, experimental animals never moan, scream, cry, growl, whimper, howl, snarl or whine; they just discreetly vocalise. Similarly, they seldom do anything so vulgar as getting killed; to the contrary, they are politely sacrificed—a term that combines a sense of devout awe at the importance of the project with an urbane sense of the scientist's reluctance to proceed to such gross courses. This theme is dealt with in more detail in Gill Langley's essay in this volume.

Many other desensitising cues serve to inculcate the same attitudes in a way somewhat like the kind of hardening that medical students necessarily undergo—a conscious suppression of normal sensibilities. There are, however, interesting differences. For medical students, it is well-understood that the hardening must be only against superficial disgusts about the appearance of blood, slime, etc.; it must not produce callousness towards another creature. Over experimental animals, it is by no means so clear that this is true. Again, the medical students' training is supposed to produce attitudes that apply to the whole of the human race; any human patient is expected to receive the same respectful and compassionate treatment. But in dealing with other species, striking anomalies appear. A few selected individuals get similar consideration, while others are treated with little or none, being so far as possible approximated to things.

The visiting scientists from another planet whom I mentioned at the beginning might be surprised at this, and might ask what determines the difference. Does it (they would wonder) depend on ethological observations about the nature of the beasts themselves, on their varying capacities for various kinds of enjoyment and suffering? The answer would be, 'Well no, actually it just depends on whether we happen to have chosen these particular animals as friends or not'. This decision is in no way scientific; it is purely social and emotional, and a lot of it is pure chance. It seems also to be very ancient, and the custom of choosing and cherishing some such animal friends is found in a great range of human societies, as James Serpell (1986) has fascinatingly shown. Scientists, like other people, usually keep the two categories sharply distinct. Not many of them would even want to imitate the great physiologist Claude Bernard, who fistulated his wife's domestic dog without warning, any more than they would calmly take their children's rabbits to cook for supper. And it is interesting to note that laboratory technicians sometimes pick out a particular mouse or mice to keep as pets, viewing them quite differently from their mass of relations in the metal cages. The same distinctness is most interestingly shown in the horror expressed in the Biblical story of the rich man who took away and cooked the poor man's one ewe lamb.

There is, too, a whole group of scientists—vets—whose work normally involves taking the personal, considerate approach to non-human creatures, because their clients are already doing so. But the two approaches cannot really be kept distinct without mutual interference, any more than they could in the well-known case of human slavery. Many situations bring them sharply into practical conflict, notably those which affect the vets themselves in relation to modern industrial methods of stock-keeping, and also about experimental animals. Vets have therefore begun to be active in the current movement to study and reform conditions in these areas, notably in shaping the British Animals (Scientific Procedures) Act 1986. These vets are among those who are beginning to find that they can no longer combine two such diverse systems as their normal humane attitude and the perverse behaviourist approach which regards animals—or even certain selected animals—purely as things, excluded by simple arbitrary fiat from the moral community.

This old approach is well illustrated by what has happened to the guinea-pig. The experimental use of these South American cavies has been so common that their very name has come to denote it. We speak with horror of a person being 'used as a guinea-pig' for some experimental purpose such as testing radiation effects, without even remembering that there are actual guinea-pigs, capable of living lives of their own, which are treated in this kind of way as a matter of course. If on some distant planet human beings who had arrived there were found to be a specially convenient experimental animal and were bred for that purpose, the word for 'human being' in the local language might well, after a time, come to have the same meaning. And of course, if the human beings complained, the scientists there might well make the same excuse that is likely to be made here in the case of the terrestrial cavies—namely, that they were not by their standards very large animals, nor indeed particularly intelligent ones.

Pigs are interesting too. They are lively and intelligent creatures. People who have tamed feral pigs in New Zealand have found them about as bright as dogs, and quite as active. Pigs made the mistake, however, of being the sacred animal of Baal, which gave the Hebrews a bad opinion of them, and has done them a lot of harm ever since. In this country, too, they acquired the servile and somewhat disreputable image that results from close confinement on a farm. Having nothing to do but eat, and happening to do it noisily, they were deemed to be greedy; having no room to be clean they were considered to be dirty. Recently they have become still more closely confined in batteries, which is likely to intensify these traits. In these circumstances, nobody is likely to notice their behaviour patterns except insofar as they cause practical inconvenience, and the image merely becomes more and more stereotyped. Though they are not yet specially prominent as experimental animals, pigs are of interest here because, like rats, they are another glaring case of an animal which is treated without consideration because it is thought of as a personified vice—an attitude which, whatever else may be said of it, certainly is not scientific.

Is there however—as the customs of our culture still make us wonder—something foolish and monstrous about the whole suggestion that we ought to treat rats and guinea-pigs with some consideration? There may be cultures where such a suggestion could not be understood at all, especially where the animal is urgently needed for food. But ours is not really one of them. Humane values are central to our official morality. At times, therefore, we see the creatures' objection to ill-treatment quite plainly, and if (for instance) our children were to start cutting them up for fun, we would interfere. Similarly, if other intelligent beings were to start cutting us up, we should probably think that, apart from merely disliking it, we had a serious grievance against them, which we would try to state. We do not really put this issue right outside morality; we simply find it very confusing, and therefore handle it (as we do other doubtful cases) by avoiding thinking about it as much as possible. Scientific use of animals is now held to need justification, which is provided partly on grounds of human benefit and partly by stressing the value of knowledge. These, however, are not viewed as all-purpose defences; if either the use or the knowlege is trivial, the justification vanishes (Midgley 1981b).

CONCLUSIONS

This is not an easy topic. Nevertheless, it is possible to think about it, and I would like to end by simply urging that we should try to do so, that we do not let ourselves be guided simply by what is familiar. Traditional thought on such painful points is bound to be confused, and what has been advanced as a scientific replacement for it has often been no better, sometimes worse. The moral community to which we take ourselves to belong is not a clear, fixed one; it has varying and shadowy boundaries. The differences between our species and those around us are not simple and definite, but extremely complex and obscure. We are not the only unique species. Elephants, as much as ourselves, are in many ways unique; so are albatrosses; so are giant pandas. All serious study of the peculiarities of any species ought to send us back to the drawing-board.

REFERENCES

Boakes, R. (1984). *From Darwin to Behaviourism, Psychology and the Minds of Animals*, Cambridge University Press, Cambridge

Darwin, C. (1872). *The Expression of Emotions in Man and Animals*, John Murray, London

Griffin, D. R. (1984). *Animal Thinking*, Harvard University Press, Cambridge, MA

Humphrey, N. K. (1976). In Bateson, P. P. G. and Hinde, R. A. (eds), *Growing Points in Ethology*, Cambridge University Press, Cambridge

Humphrey, N. K. (1978). Nature's psychologists, *New Sci.*, **78**, 900–3

Midgley, M. (1978). *Beast and Man, the Roots of Human Nature*, Harvester, Brighton

Midgley, M. (1981a). *Heart and Mind—the Varieties of Moral Experience*, Harvester, Brighton

Midgley, M. (1981b). In Sperlinger, D. (ed.), *Animals in Research—New Perspectives in Animal Experimentation*, John Wiley, Chichester, pp. 319–36

Midgley, M. (1984). *Wickedness, A Philosophical Essay*, Routledge & Kegan Paul, London

Rupke, N. A. (1987). *Vivisection in Historical Perspective*, Croom Helm, London, p. 27

Serpell, J. (1986). *In the Company of Animals—a Study of Human–Animal Relationships*, Basil Blackwell, Oxford

Stephens, M. L. (1986). *Maternal Deprivation—Experiments in Psychology, a Critique of Animal Models*, A report of the American Anti-Vivisection Society

2. Ill-gotten Gains*

Tom Regan

'I hold that, the more helpless a creature, the more entitled it is to protection by man from the cruelty of man.'

Mohandas Karamchand Ghandi (1927).
In *An Autobiography* (1982), Penguin, Harmondsworth

THE STORY

Late in 1981 a reporter for a large newspaper (we'll call her Karen to protect her interest in remaining anonymous) gained access to some previously classified government files. Karen was investigating the government's funding of research into the short- and long-term effects of exposure to radioactive waste. It was with understandable surprise that, included in these files, she discovered the records of a series of experiments involving the induction and treatment of coronary thrombosis (heart attack). Conducted over a period of 15 years by a renowned heart specialist (we'll call him Dr Ventricle) and financed with government funds, the experiments in all likelihood would have remained unknown to anyone outside Dr Ventricle's sphere of power and influence had not Karen chanced upon them.

Karen's surprise soon gave way to shock and disbelief. In case after case she read of how Ventricle and his associates took otherwise healthy individuals, with no previous record of heart disease, and intentionally caused their heart to fail. The methods used to occasion the 'attack' were a veritable shopping list of experimental techniques, from massive doses of stimulants (adrenaline was a favourite) to electrical damage of the coronary artery, which, in its weakened state, yielded the desired thrombosis. Members of Ventricle's team then set to work testing the efficacy of various drugs developed in the hope that they would help the heart withstand a second 'attack.' Dosages varied, and there were the usual control groups. In some cases, certain drugs administered to 'patients' proved more efficacious than cases in which others received no medication or smaller amounts of the same drugs. The research

*This essay first appeared in VanDeVeer, D. and Regan, T. (eds) (1987). *Health Care Ethics: An Introduction*, Temple University Press, Philadelphia, and is reprinted here [with minor amendments] with the kind permission of the publishers.

came to an abrupt end in the autumn of 1981, but not because the project was judged unpromising or because someone raised a hue and cry about the ethics involved. Like so much else in the world at that time, Ventricle's project was a casualty of austere economic times. There simply wasn't enough government money available to renew the grant application.

One would have to forsake all the instincts of a reporter to let the story end there. Karen persevered and, under false pretences, secured an interview with Ventricle. When she revealed that she had gained access to the file, knew in detail the largely fruitless research conducted over 15 years, and was incensed about his work, Ventricle was dumbfounded. But not because Karen had unearthed the file. And not even because it was filed where it was (a 'clerical error', he assured her). What surprised Ventricle was that anyone would think there was a serious ethical question to be raised about what he had done. Karen's notes of their conversation include the following:

Ventricle: But I don't understand what you're getting at. Surely you know that heart disease is the leading cause of death. How can there be any ethical question about developing drugs which *literally* promise to be life-saving?

Karen: Some people might agree that the goal—to save life—is a good, a noble end, and still question the means used to achieve it. Your 'patients', after all, had no previous history of heart disease. *They* were healthy before you got your hands on them.

Ventricle: But medical progress simply isn't possible if we wait for people to get sick and then see what works. There are too many variables, too much beyond our control and comprehension, if we try to do our medical research in a clinical setting. The history of medicine shows how hopeless that approach is.

Karen: And I read, too, that upon completion of the experiment, assuming that the 'patient' didn't die in the process—it says that those who survived were 'sacrificed'. You mean killed?

Ventricle: Yes, that's right. But always painlessly, always painlessly. And the body went immediately to the lab, where further tests were done. Nothing was wasted.

Karen: And it didn't bother you—I mean, you didn't ever ask yourself whether what you were doing was wrong? I mean . . .

Ventricle (interrupting): My dear young lady, you make it seem as if I'm some kind of moral monster. I work for the benefit of humanity, and I have achieved some small success, I hope you will agree. Those who raise cries of wrongdoing about what I've done are well-intentioned but misguided. After all, I use animals in my research—primates, to be more precise—not human beings.

THE POINT

The story about Karen and Dr Ventricle is just that—a story, a small piece of fiction. There is no real Dr Ventricle, no real Karen, and so on. But there *is* widespread use of animals in scientific research, including research like our imaginary Dr Ventricle's. So the story, while its details are imaginary—while it is, let it be clear, a literary device, not a factual account—is a story with a point. Most people reading it would be morally outraged if there actually were a Dr Ventricle who did coronary research of the sort described on otherwise healthy human beings. Considerably fewer would raise a morally quizzical eyebrow when informed of such research done on animals, primates or whatever. The story has a point, or so I hope, because, catching us off-guard, it brings this difference home to us, gives it life in our experience, and, in doing so, reveals something about ourselves, something about our own constellation of values. If we think what Ventricle did would be wrong if done to human beings but all right if done to primates, then we must believe that there are different moral standards that apply to how we may treat the two—human beings and other primates. But to acknowledge this difference, if acknowledge it we do, is only the beginning, not the end, of our moral thinking. We can meet the challenge to think well from the moral point of view only if we are able to cite a *morally relevant difference* between humans and non-human primates, one that illuminates in a clear, coherent and rationally defensible way why it would be wrong to use humans, but not primates, in research like Dr Ventricle's.

THE LARGER CONTEXT

That we cannot rationally avoid this challenge is an idea that has only recently taken root in some quarters. Cora Diamond, a philosopher at the University of Virginia, notes that a recent bibliography on society, ethics and the life sciences was described by its publishers as 'containing the most pertinent references on precisely such subjects as experimentation, containing nine pages on ethical and legal problems of experimentation, including, besides general material, sections specifically on experimentation on fetuses, prisoners, mental patients and children'; the work includes *no* references to 'the ethical problems of animal experimentation' (Diamond 1981). The explanation of this omission cannot be that there was, at that time, no (or not enough) literature on the topic, as even a cursory glance at Charles R. Magel's (1981) *A Bibliography on Animal Rights and Related Matters* will reveal. By far the likelier explanation, as Diamond observes, is that 'for many working in the field, the phrases "ethical problems posed by research" and "ethical problems posed by research *on human subjects*" are treated as simply interchangeable' (Diamond 1981). To treat them so is not to meet the challenge to give a clear, coherent and rationally defensible basis for allowing research on animals that we would not allow on humans. It is, instead, symptomatic of the moral prejudices of those who persist in assuming that there is no ethical challenge to be met.

Because most animal experiments are claimed to have medical relevance, this essay examines only the ethics of the use of animals in medical research. It is well to remind ourselves, however, of the magnitude and variety of animal use in scientific settings generally. Estimates of the total number of animals used for scientific purposes vary, some placing the total for the USA alone between 20 and 40 million, others as high as 100 million, just for a single year. In Britain, about 3 million animals are used for experimental purposes and considerably more when routine scientific procedures are included. Worldwide, the total is in the region of 250 million. Of this total, perhaps about a quarter are used in medical research, given any uncontorted meaning of the expression 'medical research' and allowing, as before, that estimates vary and are difficult to verify with anything approaching certainty. The remaining animals are used in toxicity testing (in which 'animal models' are used to estimate the risks and levels of harm humans are likely to run by using, or by being exposed to, the ever-increasing array of therapeutic and non-therapeutic products, from oven cleaners to eyeshadow, from asbestos to interferon), for basic studies of body structure and function, and in other scientific contexts.

We would serve our purposes ill, moreover, if we failed to remind ourselves of the variety of research that falls within the category of medical research as well as the multiplicity of means used to conduct it. Burn experiments (immersion of a part or the whole of an animal's body in boiling water, use of hot plates and blow torches); the infliction of tumours; radiation research (animals are studied after exposure to both small and large levels of radioactivity, an ongoing type of 'medical research' conducted in connection with 'defence' experiments by the military); the breaking of bones and bruising of flesh as a preliminary to the study of traumatic shock; brain research (cats and primates are favourite test animals, with drugs, electrodes and surgical alterations, for example, used to influence and manipulate behaviour—these are a sample of the types of experiments and methods current in medical research. When the scope and intent of the research are more psychological than physical, the methods employed vary accordingly. Research using 'aversive stimuli' commonly involves electrical shocks administered to the feet or tails of, for example, rabbits, guinea-pigs,

rats, or mice; immobilisation research (for example, dogs are sus-
pended in so-called Pavlovian slings or primates are strapped in
restraining chairs); blinding and other investigations of sensory
deprivation (for example, on monkeys in the course of studying
emotional development); aggression research (here test animals are
induced by researchers to fight among themselves); stress experiments
(any and all of the above methods, or loud noises, or random blasts of
air can be used to produce stress, the effects of which may then be
studied scientifically)—these (and there are many more) approaches
are illustrative of psychological research and methods involving
animals.

No doubt some will deny the propriety of including some of the
foregoing in the general category of medical research. These sorts of
disagreements are to be expected. However they are resolved, the
differences in the methods used in medical research, as well as the
differences in the specific form such research takes, should not obscure
their similarities. All such research, we may assume, has as its goals the
advancement of human knowledge and the improvement of public
health. These are laudatory ends. Our interest in what follows lies in
morally assessing some of the means used to achieve them. Our own
'moral research' will use Dr Ventricle's work on primates as its 'model'.

LEGAL STATUS AND THE 'RIGHT' SPECIES

Among the differences between humans and other primates, one
concerns their legal standing. It is against the law to do to human beings
what Ventricle did to his primates. It is not against the law to do this to
animals. So, here we have a difference. But a morally relevant one?

The difference in the legal status of monkeys and humans would be
morally relevant if we had good reason to believe that what is legal and
what is moral go hand in glove: where we have the former, there we
have the latter (and maybe vice versa too). But a moment's reflection
shows how bad the fit between legality and morality sometimes is. A
century and a half ago, the legal status of black people in the United
States was similar to the legal status of a house, corn, a barn: they were
property, other people's property, and could legally be bought and sold
without regard to their personal interests. But the legality of the slave
trade did not make it moral, any more than the law against drinking,
during the era of that 'great experiment' of Prohibition, made it
immoral to drink. Sometimes, it is true, what the law declares illegal
(for example, murder and rape) is immoral, and vice versa. But there is
no necessary connection, no pre-established harmony between
morality and the law. So, yes, the legal status of monkeys and humans

differs; but that does not show that their *moral* status does. Their difference in legal status, in other words, is not a morally relevant difference and will not morally justify using these animals, but not humans, in Ventricle's research.

An obvious difference, one that is biological, not legal, is that non-human primates and humans belong to different species. Once more, a difference certainly; but a morally relevant one? Suppose, for the sake of argument, that a difference in species membership *is* a morally relevant difference. If it is, and if A and B belong to two different species, then it is quite possible that killing or otherwise harming A is wrong, while doing the same things to B are not.

Let us test this idea by imagining that Steven Spielberg's E.T. and some of E.T.'s friends show up on Earth. Whatever else we may want to say of them, we do not want to say that they are members of our species, the species *Homo sapiens*. Now, if a difference in species is a morally relevant difference, we should be willing to say that it is *not* wrong to kill or otherwise harm E.T. and the other members of his biological species by hunting for sport, for example, even though it *is* wrong to do this to members of our species for this reason. But no double standards are allowed. If *their* belonging to a different species makes it all right to kill or harm them, then *our* belonging to a different species than the one to which they belong will cancel the wrongness of their killing or harming us. 'Sorry, chum', E.T.'s compatriots say, before taking aim at us or prior to inducing *our* heart attacks, 'but you just don't belong to the right species'. As for us, we cannot lodge a whine of a moral objection if species membership, besides being a biological difference, is a morally relevant one. Before we give our assent to this idea, therefore, we ought to consider whether, were we to come face to face with another powerful species of extraterrestrials, we would think it reasonable to try to move them by the force of moral argument and persuasion. If we do, we will reject the view that species differences, like other biological differences (e.g. race or sex), constitute a morally relevant difference of the kind we seek. But we will also need to remind ourselves that no double standards are allowed: though non-human primates and humans do differ in terms of the species to which each belongs, that difference by itself is not a morally relevant one. Ventricle could not, that is, defend his use of animals rather than humans in his research on the grounds that they belong to a different species from our own.

THE SOUL

Many people evidently believe that theological differences separate humans from other animals. God, they say, has given us immortal souls. Our earthly life is not our only life. Beyond the grave there is eternal life—for some, heaven; for others, hell. Animals, alas, have no soul, in this view, and therefore have no life after death either. That, it might be claimed, is the morally relevant difference between them and us, and that is why, so it might be inferred, it would be wrong to use humans in Ventricle's research but not wrong to use animals.

Only three points will be urged against this position here. First, the theology just sketched (*very* crudely) is not the only one competing for our informed assent, and some of the others (most notably, religions from the East and those of many native American peoples) do ascribe soul and an afterlife to animals. So before one could reasonably use this alleged theological difference between humans and animals as a morally relevant difference, one would have to defend one's theological views against theological competitors. To explore these matters is well beyond the limited reach of this essay. It is enough for our purposes to be mindful that there is much to explore.

Secondly, even assuming that humans have souls, while animals lack them, there is no obvious logical connection between these 'facts' and the judgement that it would be wrong to do some things to humans that it would not be wrong to do to animals. Having (or not having) a soul obviously makes a difference concerning the chances that one's soul will live on. If primates lack souls, their chances are nil. But why does that make it quite all right to use them *in this life* in Ventricle's research? And why does our having a soul, assuming we do, make it wrong *in this life* to use us? Many more questions are avoided than addressed by those who rely on a supposed theological difference between humans and animals as their basis for judging how each may be treated.

But thirdly, and finally, to make a particular theology the yardstick of what is permissible and, indeed, supported by public funds in a pluralistic society such as we find in the 20th century Western world is itself morally objectionable, offending, minimally, the sound moral, not to mention legal, principle that Church and State be kept separate. Even if it had been shown to be true, which it has not, that humans have souls and animals do not, that should not be used as a weapon for making public policy. We will not, in short, find the morally relevant difference we seek if we look for it within the labyrinth of alternative theologies.

THE RIGHT TO CONSENT

'Human beings can give or withhold their informed consent; animals cannot. That's the morally relevant difference.' This argument is certainly mistaken on one count, and possibly mistaken on another.

Concerning the latter point first, evidence steadily increases regarding the intellectual abilities of chimps and other primates (e.g. gorillas). Much of the public's attention has been focused on reports of studies involving the alleged linguistic abilities of these animals, when instructed in such languages as American Sign Language for the deaf (ASL). Washoe. Lana. Nim Chimpski. Individual chimps have attained international notoriety. But how much these animals do and can understand is very much up in the air at this point. Whether primates have sufficient ability to understand and use language and, if they do, whether they have sufficient ability to give or withhold their informed consent—these matters cannot be settled arbitrarily at this point in time. Possibly these animals lack these abilities. But possibly they do not. Those who trot out a doctrinaire position in this regard prove how little, not how much, they know.

Questions about the ability of chimps to give informed consent aside, it should be obvious that this ability is not the morally relevant difference we are seeking. Suppose that, in addition to using primates, Ventricle also used some humans, but only mentally incompetent ones—those who, though they have discernible preferences, are too young or too old, too enfeebled or too confused, to give or withhold their informed consent. If the ability to give or withhold informed consent were the morally relevant difference we seek, we should be willing to say that it was not wrong for Ventricle to do his coronary research on these humans, though it would be wrong for him to do it on competent humans—those humans, in other words, who can give or withhold their informed consent.

But though one's willingness to consent to have someone do something to oneself may be, and frequently is, a good reason to absolve the other person of moral responsibility, one's inability to give or withhold informed consent is on a totally different moral footing. When Walter Reed's colleagues gave their informed consent to take part in the yellow fever experiments, those who exposed them to the potentially fatal bite of the fever parasite carried by mosquitoes were absolved of any moral responsibility for the risks the volunteers chose to run, and those who chose to run these risks, let us agree, acted above and beyond the normal call of duty—acted, as philosophers say, supererogatorily. Because they did more than duty strictly requires, in the hope and with the intention of benefiting others, these pioneers deserve our esteem and applause.

The case of human incompetents is radically different. Since these humans (e.g. young children and the mentally retarded) lack the requisite mental abilities to have duties in the first place, it is absurd to think of them as capable of acting supererogatorily; they cannot act 'beyond the call' of duty, when, as is true in their case, they cannot understand that 'call' to begin with. But though they cannot volunteer, in the way mentally competent humans can, they can be forced or coerced to do something against their will or contrary to their known preferences. Sometimes, no doubt, coercive intervention in their life is above moral reproach—indeed, is morally required, as when, for example, we force a young child to undergo a spinal tap to check for meningitis. But the range of cases in which we are morally permitted or obliged to use force or coercion on human incompetents in order to accomplish certain ends is not large by any means. Primarily it includes cases in which we act with the intention, and because we are motivated, *to forward the interests of that individual human being.* And that is not a licence, not a blank cheque to force or coerce human incompetents to be put at risk of serious harm so that *others* might possibly be benefited by having *their* risks established or minimised. To treat the naturally occurring heart ailment of a human incompetent *is* morally imperative, and anything we learn as a result that is beneficial to others is not evil by any means. However, intentionally to bring about the heart attack of a human incompetent, on the chance that others might benefit, is morally out of bounds. Human incompetents do not exist as medical resources for the rest of us. Morally, Ventricle's research should be condemned if done on human incompetents, whatever benefits others might secure as a result. Imagine our gains to be as rich and real as you like. They would all be ill-gotten.

What is true in the case of human incompetents (those humans, once again, who, though they have known preferences, cannot give or withhold their informed consent) is true of primates (and other animals like them in the relevant respects, assuming, as we are, that animals cannot give or withhold their informed consent). Just as in the case of these humans, so also in the case of these animals, we are morally permitted and sometimes required to act in ways that coercively put them at risk of serious harm, against their known preferences, as when, for example, they are subjected to painful exploratory surgery. But the range of cases in which we are justified in using force or coercion on them is morally circumscribed. Primarily it is to promote *their* individual interests, as we perceive what is in their interests. It is *not* to promote the collective interests of *others*, including those of human beings. Animals are not our tasters, we are not their kings. To treat them in ways that put them at risk of significant harm on the chance that we might learn something useful, something that just might add to our understanding of disease or its treatment or prevention—coercively to put them at risk of significant harm for any or all of these reasons—is morally to be condemned.

To attempt to avoid this finding in the case of these animals, while holding on to the companion finding in the case of incompetent humans, is as rational as trying to whistle without using your mouth. It can't be done. As certain as it is that it would have been wrong for Ventricle to use human incompetents in his coronary research, it is at least as certain that it would have been wrong for him to use primates instead, despite the legality of using these animals and the illegality of using these humans, and notwithstanding the actual biological and alleged theological differences between these humans and animals. Whatever gains we might have harvested, for present or future generations of human beings, would have been ill-gotten.

THE INDIRECT DUTY RESPONSE

People try to avoid this conclusion in a variety of ways. For example, some argue that we do not have any duties *to* animals (what philosophers call direct duties); rather, we have only duties *involving* animals (so-called indirect duties). Animals, in this view, have the same kind of moral status as redwoods, the Taj Mahal and El Greco's *View of Toledo*. Few would deny that we have a duty to preserve these things, but most would deny that we have a duty to *them* to do so. Our duty, most people seem to think, is a duty *to other human beings*, both present and future generations, to preserve great works of art and the majesty of nature so that they, these other human beings, might have an opportunity to see and appreciate them, thereby enriching the quality of their lives. Duties involving works of art and the majesty of nature, in short, are indirect duties to humanity.

The same is true, some people maintain, of our duties regarding animals. By all means, don't harm them unnecessarily, they say; but don't be misled into thinking that this is because we have duties directly to them. When animals are owned by others, we certainly ought not to harm them unnecessarily because, after all, we have a duty to property owners not to harm their property. And when animals are not owned by anyone in particular, we still should not harm them unnecessarily since people who do this to animals have a tendency to do the same sorts of things to human beings; since we *do* have a duty not to do this sort of thing to human beings, we therefore ought to avoid doing it to animals—not because we owe it to them, to be sure, but because we do owe it to one another.

If our duties regarding animals were indirect duties, one might then argue that Ventricle's research would have been wrong if done on human incompetents, but morally permissible if done on primates. If our duties regarding animals are indirect duties to humanity, then the morality of how we treat them is to be decided by what promotes human interests, and it is certainly possible that our interests would be promoted more by allowing animal research like Ventricle's than if we banned it.

But what about human incompetents? What type of duty do we have in their regard—direct or indirect? If one affirms direct duties in their case, while denying direct duties owed to other animals, then, once again, one will want to be told what is the morally relevant difference between these humans and animals, something that, as we know, has not been established by the arguments so far considered. Moreover, one cannot say that the morally relevant difference simply is that the duties owed to human incompetents are direct, while those involving animals are indirect, since this view presupposes that a morally relevant difference exists between the two. This view, therefore, cannot itself specify what that difference is.

The second option is to hold that our duties involving human incompetents, like our duties regarding animals, are indirect. This option at least has the merit of being consistent. Its principal defect is that it is false. Morally, it is preposterous to maintain that the reason why we ought not torture little children, for example, or kill their senile grandparents is because of the interests of others—for example, the children's parents or other elderly, more lucid people who, learning of the fate that befalls the senile, will live out their last years in wretched anxiety. It *is* wrong to torture children. But it is wrong to do this because in doing it we violate a duty we have directly to individual children *quite apart* from what their parents (or anyone else) happen to think or feel. And the same is true in the case of other harms we might visit upon other human incompetents. We owe it to them directly not to harm them. If others benefit in the bargain when we do as duty requires, they may count themselves lucky. But whether or not others benefit as a result of our refusal to harm human incompetents is, strictly speaking, morally irrelevant to whether we have duties to them. Our duties regarding human incompetents are not indirect duties to other people.

Of the two options, therefore, the second (that we have direct duties to human incompetents) is the one we should accept. Not to do so would be to distort, rather than illuminate, the moral status of these humans. Once more, however, we cannot consistently regard the moral status of animals any differently than that of these humans, if we are unable to cite a morally relevant difference. In other words, since our duties regarding human incompetents are direct duties, since we have duties regarding animals and assuming we are unable to cite and defend a morally relevant difference between these animals and humans, then our duties regarding animals are likewise direct, not indirect, duties—*duties we owe directly to them, considered as individuals*. In particular, therefore, we owe it to these animals themselves not to harm them. Any further moral thinking about these animals must both take this into account and be able to account for it, and any treatment of animals that rests on a view about these animals that is deficient in these respects cannot be rationally satisfactory. One cannot, therefore, defend the gains others might have received from Ventricle's research on primates, but condemn any gains stemming from such research if done on human incompetents, by claiming that our duties to these animals are indirect, while those involving these humans are direct. If one end is ill-gotten, then so is the other.

THE CONTRACTARIAN RESPONSE

Contractarianism is a second position that might seem to support research on animals but not on incompetent humans. Roughly speaking, contractarians view morality as consisting of a set of mutually agreed constraints on everyone's behaviour. Each party to these agreements (or 'the contract'), we are to suppose, seeks to maximise what is in his or her individual self-interest. Each party soon realises, however, that, to achieve this objective, others must be limited in what they may do. For example, it is self-defeating for Friday to work to secure food and a place to live if Crusoe is at liberty to steal his property. Since no-one has any self-interested reason to limit the pursuit of his or her self-interest unilaterally, such limits can come into being only if enough people agree to abide by them and, relatedly, agree to impose appropriate sanctions (e.g. fines or other punishments) on those who fail to cooperate.

Contractarians have important intramural differences. Some believe that present-day morality can be traced to an actual historical agreement ('the original contract'); others interpret the notion of a contract ahistorically. These and other internal differences to one side, it should be clear that animals, primates included, can find a precarious home at best within standard versions of contractarianism. As far as we know, human beings are the only terrestrial creatures capable of entering into contracts. That being so, what duties, if any, we have regarding animals must depend on what these human contractors judge to be in *their* (human) self-interest. If most of these humans agree that it is in their individual self-interest to allow Ventricle-like research on primates, while forbidding analogous research on human incompetents, then the former, but not the latter, research would be justified. In this way, then, contractarianism might seem to provide an adequate moral basis for Ventricle's research.

But contractarianism, at least given one of its expressions, could justify far more than research on animals. If enough people happen to believe that it would be in their self-interest to suppress or oppress the members of a given minority (e.g. a racial or religious minority), when it comes to such vital matters as access to medical care, education, or career opportunities, then such policies, if mutually agreed upon by the majority, could not be morally condemned, given this version of contractarianism. That approach to morality, in other words, has the undesirable feature of legitimating the philosophy that might, understood as the collective judgement and power of the majority, makes right. Few, if any, will find this a congenial moral philosophy, since it would justify the most extreme expressions of racial and other forms of oppression. To have recourse to this philosophy as a defence of Ventricle-type research on primates, therefore, is like trying to keep one's moral position afloat by drilling a hole in it. Our (human) might does give us the power to use animals in research, just as the might of the majority gives it the power to exploit the members of racial or religious minorities. But in neither case does might make right.

One could, it is true, endeavour to retain the spirit of contractarianism while altering the letter somewhat. Instead of allowing the parties to the contract to know, for example, their race, sex, religion and nationality, one might ask them to imagine that they stand 'behind a veil of ignorance', a veil that is thick enough to preclude their knowing the particular details of their life, thereby ensuring that they will select principles of justice impartially rather than on narrow, partisan grounds. Such a view can be found in John Rawls's influential work, *A Theory of Justice* (Rawls 1971). This is not the occasion to offer a full account or lengthy assessment of Rawls's version of contractarianism. Here it must suffice to note that Rawls, while insisting that the veil of ignorance keeps his contractors in the dark about, for example, what race or sex they will be, allows them to know that they will be members of the human race. Small wonder, then, that Rawls's view implies that we do have duties directly to one another but not to animals. The cards, as dealt by Rawls, are stacked against ensuring an impartial judgement of the moral status of animals. Though his theory has much to recommend it, we will not find in it a rationally satisfying basis for defending research like Ventricle's.

THE UTILITARIAN RESPONSE

A third view worthy of consideration concedes that animals are on all fours, so to speak, with human incompetents, when it comes to their respective moral status: both are owed the same basic duty owed to those humans who are competent (that is, who have the ability to give or withhold their informed consent and, in having this ability, have all those other cognitive abilities thereby presupposed). That duty is twofold: first, to consider their interests and, secondly, to count equal interests equally. The interests in question are what individuals prefer or would rather do without—what they like or dislike, love or hate, what they are 'for' or 'against'. In the case of every individual with interests, then, we must first take the time and trouble to ask what his or her interests are before we can decide what, morally speaking, we ought to do. More than this, however, we must also weigh or count equal interests equally. If Jack and Jill both want to go up the hill, and if they both want to do so equally, then we must count their interests in going up the hill as being equal in importance. Rationally, we cannot discount the importance of Jill's interest on the grounds that 'she's only a girl' or Jack's because 'he's a dumb boy'. To treat Jack and Jill fairly, to treat them equitably, to treat them as equals, requires that we consider their respective interests and count their like interests as of like importance. Let us refer to the principle that demands equal consideration and weighting of like interests as the *equality of interests principle* or *equality principle*.

The equality of interests principle is one part of a currently fashionable view called 'preference utilitarianism'. Utilitarianism, very roughly speaking, requires that we act in order to bring about the optimum aggregate balance of good over bad consequences for all those affected by what we do. If we think of 'the good' as involving the satisfaction of individual preferences, and 'the bad' as involving the frustration of such preferences, then the close connection between the principles of utility and equality should be clear. As preference utilitarians, what we aim to bring about is the best aggregate balance of the satisfaction of preferences over their frustration for all affected by what we do. To aim at this objective, however, we must first consider who has what preferences (interests) and count equal interests equally; we must, that is, first rely on the equality of interests principle.

Some preference utilitarians think that preference utilitarianism would condemn Ventricle's research on primates. The arguments these thinkers give, however, are far from convincing. Essentially, what they come to is the claim that allowing research on the animals while forbidding it on human incompetents must violate the equality of interests principle, a principle that, given preference utilitarianism, it is always wrong to violate. But those who argue in this way are confused. Suppose both Clint Eastwood and Clyde (the orang-utan in the film *Any Which Way You Can*) have the same interest in avoiding the excruciating pain associated with a heart attack. As preference utilitarians, we certainly must take the interests of both into account, and, assuming their equality, we must count them as being of equal importance. It does not follow from our having done this, however, that we must now approve of doing only those things to Clint that we would approve of doing to Clyde, and vice versa. What we ought to do to either, assuming we have observed the strictures of the equality principle, is now to be determined by appeal to the principle of utility. And there is no reason why *the consequences for others* (namely all those who will be affected by the consequences of our acts) will be the same if we do only the same things to Clyde as we would be willing to do to Clint. In particular, it is certainly possible that the aggregate balance of good over bad for all affected by the outcome would be better if Ventricle did his research on Clyde than if he did it on Clint.

It does not follow from this that preference utilitarians cannot condemn research such as Ventricle's when done on primates. What follows is, first, that they cannot condemn it on the ground that it must violate the equality of interests principle (for it need not), and, secondly, that they must acknowledge that whether or not they have grounds on which to condemn it depends on their having knowledge of the relevant consequences of Ventricle-like research. What consequences, then, are relevant? In the nature of the case, these must deal with the degree to which the interests (preferences) of all those affected by the outcome of the research are satisfied or frustrated. This requires more than our knowing how Clyde's interests would fare. There are also Ventricle's interests to take into account, as well as those of his staff, plus those who build the tools of the medical researcher's trade (e.g. cages, restraint chairs), plus those who have an economic interest in the development of new drugs, plus our vital interest in health, and so on. There are, in a word, numerous interests to take into account and assess equitably before anyone could plausibly claim, with any degree of credibility, that the consequences of not allowing Ventricle to do his research on primates would not bring about the best aggregate balance of preference satisfaction over frustration for all those affected by the outcome. Indeed, given that what we are being asked to do is compare the interests of relatively few animals against the not unimportant (e.g. economic, scientific and health) interests of many more human beings, the utilitarian case seems to bode ill for the primates. On the face of it, there is a very strong presumptive utilitarian case to be made in favour of Ventricle's research when done on primates.

This might seem to be good news for those who favour such research. It isn't. Preference utilitarianism does more than offer a way to justify Ventricle-type research when done on animals. It will also justify similar research done on human incompetents or, for that matter, on human competents, even without or against their informed consent. Granted, in cases involving humans, just as in cases involving animals, we must take pains to consider everyone's interests and count equal interests equally. Having done this, however, there is no reason why, in this or that case, the aggregate consequences for others might not be 'the best' if we allowed research to be done on the humans in question. If we agree that research involving these humans is wrong, then we shall certainly want a moral principle that will not sanction it. That being so, we shall certainly want to avoid preference utilitarianism. It is no rational defence of Ventricle-type research, therefore, to note that preference utilitarianism will, or very likely will, allow it. That view will, or very likely will, allow a great deal that is wrong. To show that preference utilitarianism would sanction research on animals is far from showing that such research is morally tolerable.

PERFECTIONIST AND UTILITARIAN VIEWS OF VALUE

Thus far we have advanced a controversial moral thesis—namely, that research like Ventricle's, when done on primates, cannot be justified by appealing to the benefits others do or might receive. And we have also considered, only to reject, various responses that seek to refute this thesis. Even if this thesis and its defence to this point are sound, much philosophical work remains to be done. A controversial thesis like the one before us does not stand on its own two feet. One must not only defend it against likely objections, a task to which we have attended, if incompletely, in the foregoing; one must also attempt to identify and defend the moral grounds on which the thesis stands, a task we have thus far failed to undertake. When, as now, we turn our attention to this item on our agenda, we must anticipate that the full weight of this challenge cannot be borne here.

Although preference utilitarianism is not the adequate position its advocates suppose, the emphasis it places on treating individuals as equals is an important corrective to less egalitarian visions of morality. The ancient Greek philosopher Aristotle, for example, offers a perfectionist moral theory: people are better (and so deserve more) than others if they have a certain cluster of intellectual and artistic excellences (virtues). Indeed, some people are so lacking in the favoured virtues that Aristotle thinks they are born to be the slaves of those who are more generously endowed. Perfectionism of the Aristotelian sort must strike us as morally offensive, and it is one of the virtues of utilitarianism, because of the importance it places on treating relevantly similar individuals as equals, that it disassociates itself from perfectionism.

But the *type* of equality we find in preference (and other forms of) utilitarianism is easily misunderstood. For the preference utilitarian (to limit our attention to this version of utilitarianism), it is not individuals that count as equals but rather their mental states—their preference satisfactions and frustrations. Individuals are *receptacles* of value, things into which, so to speak, what has value can be 'poured', like liquid into a cup. But it is the liquid in the cup (that is, the mental states of satisfaction or frustration) that has value, whether positive or negative. The cup (that is, the individual human being or, for that matter, the individual animal), though 'containing' what has value, has no value of its own.

To view humans or animals in this way is to offer a theory of their value (or, more precisely, their lack of it) that can legitimate using them as medical and other sorts of resources, when this theory of value is coupled with the utilitarian injunction to act in order to bring about the best aggregate balance of good over bad for all affected by the outcome. If we consider the interests of all those who will be affected, if we count equal interests equally, and if, having done this, we can bring about the best aggregate balance of good over bad for all affected by doing medical research on primates (or on human incompetents, or, indeed, on unwilling competent human beings), then our research is justified. The gains others secure, on this view, are *not* ill-gotten. To maintain, as we have, that such gains *are* ill-gotten is thus implicitly to reject preference utilitarian approaches to questions about the justification of medical research. More deeply and, for present purposes, more importantly, it is also to reject standard utilitarian theories of value. According to these theories, as noted, it is what 'goes into' the individual, what the individual 'contains'—for example, the mental state of satisfaction—that has value, not the individual. Our controversial thesis about using animals such as primates in research such as Ventricle's thus turns out to involve a different vision, neither perfectionist nor utilitarian, of the value of the individual.

THE VALUE OF THE INDIVIDUAL

This alternative vision consists of viewing certain individuals as themselves having a distinctive kind of value, what we will call 'inherent value'. This kind of value is not the same as, is not reducible to and is not commensurate with either such values as preference satisfaction or frustration (that is, mental states) or such values as artistic or intellectual talents (that is, mental and other kinds of excellences or virtues). That is, we cannot equate or reduce the inherent value of an individual to his or her mental states or virtues, and neither can we intelligibly compare the two. In this respect, the three kinds of value (mental states, virtues and the inherent value of the individual) are like proverbial apples and oranges.

They are also like water and oil: they don't mix. It is not only that

Clint's inherent value is not the same as, not reducible to and not commensurate with *his* satisfaction, pleasures, intellectual and artistic skills, etc. In addition, *his* inherent value is not the same as, is not reducible to and is not commensurate with the valuable mental states or talents of *other* individuals, whether taken singly or collectively. Moreover, and as a corollary of the preceding, the individual's inherent value is in all ways independent both of his or her usefulness relative to the interest of others and of how others feel about the individual (for example, whether one is liked or admired, despised or merely tolerated). A prince and a pauper, a streetwalker and a nun, those who are loved and those who are forsaken, the genius and the retarded child, the artist and the philistine, the most generous philanthropist and the most unscrupulous used-car salesman—all have inherent value, according to the view recommended here, and all have it equally. Decidedly non-perfectionist in letter and spirit, this vision of value is decidedly non-utilitarian as well.

To view the value of individuals in this way is not an empty abstraction. To the question, 'What difference does it make whether we view individuals as having equal inherent value, or, as utilitarians do, as lacking such value, or, as perfectionists do, as having such value but to varying degree?', our response must be, 'It makes all the moral difference in the world!' Morally, we are *always* required to treat those who have inherent value in ways that display proper respect for their distinctive kind of value, and though we cannot on this occasion either articulate or defend the full range of obligations tied to this fundamental duty, we can note that we fail to show proper respect for those who have such value whenever we treat them as if they were mere receptacles of value or as if their value was dependent on, or reducible to, their possible utility relative to the interests of others. In particular, therefore, Ventricle would fail to act as duty requires—would, in other words, do what is morally wrong—if he conducted his coronary research on competent human beings, without their informed consent, on the grounds that this research just might lead to the development of drugs or surgical techniques that would benefit others. That would be to treat these human beings as mere receptacles or as mere medical resources for others, and though Ventricle might be able to do this and get away with it, and though others might benefit as a result, that would not alter the nature of the grievous wrong he would have done. And it would be wrong, not because (or only if) there were utilitarian considerations, or contractarian considerations, or perfectionist considerations against his doing his research on these human beings, but because it would mark a failure on his part to treat them with appropriate respect. To ascribe inherent value to competent human beings, then, provides us with the theoretical wherewithal to ground our moral case against using competent human beings, against their will, in research like Ventricle's.

WHO HAS INHERENT VALUE?

If inherent value could non-arbitrarily be limited to competent humans, then we would have to look elsewhere to resolve the ethical issues involved in using other individuals (for example, primates) in medical research. But inherent value can only be limited to competent human beings by having recourse to one arbitrary manoeuvre or another. Once we recognise that we have direct duties to competent and incompetent humans as well as to animals such as primates; once we recognise the challenge to give a sound theoretical basis for these duties in the case of these humans and animals; once we recognise the failure of indirect duty, contractarian and utilitarian theories of obligation; once we recognise that the inherent value of competent humans precludes using them as mere resources in such research; once we recognise that a perfectionist vision of morality, one that assigns degrees of inherent value on the basis of possession of favoured virtues, is unacceptable because of its inegalitarian implications; and once we recognise that morality simply will not tolerate double standards, then we cannot, except arbitrarily, withhold ascribing inherent value, to an equal degree, to incompetent humans and animals such as primates. All have this value, in short, and all have it equally. All considered, this is an essential part of the most adequate total vision of morality. Morally, none of those having inherent value may be used in Ventricle-like research (research that puts them at risk of significant harm in the name of securing benefits for others, whether those benefits are realised or not). And none may be used in such research because to do so is to treat them as if their value is somehow reducible to their possible utility relative to the interests of others, or as if their value is somehow reducible to their value as 'receptacles'. What contractarianism, utilitarianism and the other '-isms' discussed earlier will allow is not morally tolerable.

HURTING AND HARMING

The prohibition against research like Ventricle's, when conducted on animals such as monkeys, cannot be avoided by the use of anaesthetics or other palliatives used to eliminate or reduce suffering. Other things being equal, to cause an animal to suffer is to harm that animal—is, that is, to diminish that individual animal's welfare. But these two notions—harming on the one hand and suffering on the other—differ in important ways. An individual's welfare can be diminished independently of causing her to suffer, as when, for example, a young woman is reduced to a 'vegetable' by painlessly administering a debilitating drug to her while she sleeps. We mince words if we deny that harm has been done to her, though she suffers not. More generally, harms, understood as reductions in an individual's welfare, can take the form either of *inflictions* (gross physical suffering is the clearest example of a harm of this type) or *deprivations* (prolonged loss of physical freedom is a clear example of a harm of this kind). Not all harms hurt, in other words, just as not all hurts harm.

Viewed against the background of these ideas, an untimely death is seen to be the ultimate harm for both humans and animals, such as primates, and it is the ultimate harm for both because it is their ultimate deprivation or loss—their loss of life itself. Let the means used to kill primates be as 'humane' (a cruel word, this) as you like. That will not erase the harm that an untimely death is for these animals. True, the use of anaesthetics and other 'humane' steps lessens the wrong done to these animals, when they are 'sacrificed' in Ventricle-type research. But a lesser wrong is not a right. To do research that culminates in the 'sacrifice' of primates or that puts these and similar animals at risk of losing their life, in the hope that we might learn something that will benefit others, is morally to be condemned, however 'humane' that research may be in other respects.

THE CRITERION OF INHERENT VALUE

It remains to be asked, before concluding, what underlies the possession of inherent value. Some are tempted by the idea that life itself is inherently valuable. This view would authorise attributing inherent value to primates, for example, and so might find favour with some people who oppose using these animals in research. But this view would also authorise attributing inherent value to anything and everything that is alive, including, for example, crabgrass, lice, bacteria and cancer cells. It is exceedingly unclear, to put the point as mildly as possible, either that we have a duty to treat these things with respect or that any clear sense can be given to the idea that we do.

More plausible by far is the view that those individuals have inherent value who are *the subjects of a life*—who are, that is, the experiencing subjects of a life that fares well or ill for them over time, those who have *an individual experiential welfare*, logically independent of their utility relative to the interests or welfare of others. Competent humans are subjects of a life in this sense. But so, too, are those incompetent humans who have concerned us. And so, too, and not unimportantly, are non-human primates. Indeed, so too are the members of many species of animals: cats and dogs, monkeys and sheep, rats and rabbits, cetaceans and wolves, horses and cattle. Where one draws the line between those animals who are, and those who are not, subjects of a life is certain to be controversial. Still there is abundant reason to believe that the members of mammalian species of animals do have a psychophysical identity over time, do have an experiential life, do have an individual welfare. Common sense is on the side of viewing these animals in this way, and ordinary language is not strained in talking of them as individuals who have an experiential welfare. The behaviour of these animals, moreover, is consistent with regarding them as subjects of a life, and the implications of evolutionary theory are that there are many species of animals whose members are, like the members of the species *Homo sapiens*, experiencing subjects of a life of their own, with an individual welfare. On these grounds, then, we have very strong reason to believe, even if we lack conclusive proof, that these animals meet the subject-of-a-life criterion.

If, then, those who meet this criterion have inherent value, and have it equally relative to all who meet it, primates and other animals who are subjects of a life, not just human beings, have this value *and* have neither more nor less of it than we do. (To hold that they have less than we do is to land oneself in the inegalitarian swamp of perfectionism.) Moreover, if, as has been argued, having inherent value morally bars others from treating those who have it as mere receptacles or as mere resources for others, then any and all medical research like Ventricle's, done on these animals in the name of possibly benefiting others, stands morally condemned. And it is not only cases in which the benefits for others do not materialise that are condemnable; also to be condemned are cases, such as the research done on primates regarding hepatitis, for example, in which the benefits for others are genuine. In these cases, as in others like them in the relevant respects, the ends do not justify the means. The *many millions* of mammalian animals used each year for scientific purposes, including medical research, bear mute, tragic testimony to the narrowness of our moral vision.

CONCLUSIONS

This condemnation of such research probably is at odds with the judgement that most people would make about this issue. If we had good reason to assume that the truth always lies with what most people think, then we could look approvingly on Ventricle-like research done on animals like primates in the name of benefits for others. But we have no good reason to believe that the truth is to be measured plausibly by majority opinion, and what we know of the history of prejudice and bigotry speaks powerfully, if painfully, against this view. Only the cumulative force of informed, fair, rigorous argument can decide where the truth lies, or most likely lies, when we examine a controversial moral question. Although openly acknowledging and, indeed, insisting on the limitations of the arguments in this essay, these arguments make the case, in broad outline, against using animals such as primates in medical research such as Ventricle's. Various challenges to this position have been considered and judged inadequate, and the deeper philosophical grounds that stand beneath the surface of this controversy, grounds that concern alternative theories of the value of the individual (or lack of it), have been, if not thoroughly excavated, at least turned over. That does not bring thinking about value to an end, but it is something of a beginning.

Those who oppose the use of animals such as primates in research like Ventricle's and who accept the major themes advanced here, oppose it, then, not because they think that all such research is a waste of time and money, or because they think that it never leads to any benefits for others, or because they view those who do such research as, to use Ventricle's words, 'moral monsters', or even because they love animals. Those of us who condemn such research do so because this research is not possible except at the grave moral price of failing to show proper respect for the value of animals that are used. Since, whatever our gains, they are ill-gotten, we must bring an end to research like Ventricle's, whatever our losses. A fair measure of our moral integrity will be the extent of our resolve to work against allowing our scientific, economic, health and other interests to serve as a reason for the wrongful exploitation of members of species of animals other than our own.

REFERENCES

Diamond, C. (1981). Experimenting on animals: a problem in ethics. In Sperlinger, D. (ed.), *Animals in Research: New Perspectives in Animal Experimentation,* John Wiley, Chichester, p. 345

Magel, C. R. (1981). *A Bibliography on Animal Rights and Related Matters,* University Press of America, Washington, DC

Rawls, J. (1971). *A Theory of Justice,* Harvard University Press, Cambridge, MA

Further reading

Diner, J. (1974). *Physical and Mental Suffering in Experimental Animals,* The Animal Welfare Institute, Washington, DC

Frey, R. G. (1980). *Interests and Rights: The Case Against Animals*, Oxford University Press, Oxford

Godlovitch, S., Godlovitch, R. and Harris, J. (1972). *Animals, Men and Morals*, Taplinger, New York

Magel, C. R. (1980). In Salt, H. S. (ed.), *Animals' Rights*, Society for Animal Rights, Clarks Summit

Magel, C. R. and Regan, T. (1982). In Regan, T. (ed.), *All That Dwell Therein: Essays on Animal Rights and Environmental Ethics*, University of California Press, California

Morris, R. K. and Fox, M. W. (1978). *On the Fifth Day: Animal Rights and Human Ethics*, Acropolis Press, Washington, DC

Paterson, D. A. and Ryder, R. (1979). *Animals' Rights: A Symposium*, Centaur, Fontwell

Regan, T. (1983). *The Case for Animal Rights*, University of California Press, California

Regan, T. and Singer, P. (1976). *Animal Rights and Human Obligations*, Prentice-Hall, Englewood Cliffs, NJ

Singer, P. (1975). *Animal Liberation*, Avon, New York

Sperlinger, D. (1981). *Animals in Research: New Perspectives in Animal Experimentation*, John Wiley, Chichester

3. Evidence for Pain and Suffering in Other Animals

Margaret Rose and David Adams

'We have seen that the senses and intuitions, the various emotions and faculties, such as love, memory, attention and curiosity, imitation, reason, etc., of which man boasts, may be found in an incipient, or even sometimes in a well-developed condition, in the lower animals.'

Charles Darwin (1871). *The Descent of Man*

INTRODUCTION

Considerations of pain and suffering have been major determinants of the relationship between humankind and other animals, particularly in the development of legislative definitions of this relationship and, more recently, as the basis for the debate on the moral status of animals.

Such considerations have had a particular application to the use of animals in research. As noted by Loew (1981), the principal objection to the use of animals in research is that that research may cause pain. The primary purpose of the first legislation regulating the use of animals in research, the Cruelty to Animals Act 1876 in the United Kingdom, was to control experiments on animals which were calculated to cause pain. Since that time, a number of countries have introduced legislation to limit the circumstances under which pain and distress can be caused to animals during experiments.

There have been significant changes in emphasis in recent years. First, there has been a shift from consideration of protection from cruelty as the basis for legislation to a more broadly based concern with avoidance of pain and psychological stress (OTA 1986, Warnock 1985). Secondly, the debate about the moral status of animals has led to a re-evaluation of the justification for the use of animals in research (Tannenbaum and Rowan 1985). We now recognise that the interests of animals to be free of pain and distress may conflict with human interests in the pursuit of knowledge (Zimmerman 1984, OTA 1986). Consequently the decision to use animals depends upon a consideration of the balance of these interests (Bateson 1986, Loew 1987). This principle is now embodied in legislation in many countries, including the United Kingdom, the United States of America, Sweden, Holland, Australia and New Zealand. The avoidance and minimisation of pain and distress in animals is, increasingly, a fundamental principle guiding their use in research (Bankowski and Howard-Jones 1984). The potential for animals to experience pain and distress is the basis for consideration of their interests. Nevertheless, knowledge of pain and distress in animals and our ability to assess and treat these conditions is limited. Only recently have these phenomena been subjected to critical enquiry and debate.

An examination of *Index Veterinarius* between 1972 and 1982 reveals only 16 citations for pain during that period, but by 1986 there were 14 in one year. Further, there have been a number of texts and monographs published in the past five years specifically directed to consideration of the detection and alleviation of pain in animals (Kitchell and Erickson 1983, Iggo 1984, Moberg 1985, Gibson 1985, AVMA 1987).

Current knowledge of the biology and experience of pain has been shaped by human need and is based on human experience. Despite human interest in the relief of pain since earliest recorded history (Keele 1957), the study of pain as a biological phenomenon has been a relatively recent development. Considerations of pain and the capacity for suffering have been the basis of philosophical argument, particularly in moral philosophy, but paradoxically there is no reference to either pain or suffering in the *Encyclopaedia of Philosophy* (Edwards 1972).

Human experience of pain serves as a general model and as the paradigm for pain in animals. The subjective nature of the pain experience in people is a major obstacle to both the definition and the examination of the phenomenon, not only in human beings but, more importantly for our purposes, in animals (Kitchell *et al.* 1962).

Pain, in the human context, is not a single, specific experience which can be analysed and manipulated (Melzack 1968). Rather, it refers to a complex group of experiences which incorporate psychological and neurobiological aspects and encompass considerations of self-awareness, cognition and questions of the relationship between the mind and the body, the existence and influence of mental states and feelings. The complexity of the human model of pain engenders considerable debate about the significance of mental states and cognition in the overall experience (Rachlin 1985, Wilson 1985). Moreover, this debate has wider implications in the differing approaches to the philosophy of science, and the philosophy of the mind–body relationship in particular (Lindley *et al.* 1978). Materialistic, dualistic, functionalistic or mentalistic perspectives will determine how pain is described and investigated. The psychobiological theory developed by Bunge (1980) may assist by providing a multi-disciplinary approach.

When the human model is used for considerations of pain in animals, difficulties are compounded by questions about the existence of self-consciousness and mental states in animals. We have no empirical means of verifying that animals have conscious experiences and mental states comparable to human beings. Do animals have a mind? Higher brain functions, particularly those involved with cognition, can significantly modify pain perception (Melzack 1961). Psychological control and modification of pain is a most pervasive influence on pain perception in human beings (Elton *et al*. 1983, Holzman 1986). If the only model we have for the study of pain is the human model, how do we establish that animals experience pain, particularly when the ultimate experience in human beings is subjective and so influenced by psychological and cognitive modalities?

Zayan (1986) proposes that we establish the existence of pain in animals by analogical inference. He describes pain as being derived from two concepts which refer to two kinds of facts. One is nociception, the other feeling. Subjective pain, he contests, involves emotion or cognition, or both. This rationale for assuming that animals experience pain is based first on an analogy between nociception in human beings and animals, using comparative neuroanatomical and neurophysiological data, and then on the assumption that animals are capable of perceiving and experiencing pain, again by analogy with pain experience in humans.

Our purpose will be to describe the neural, behavioural, emotional and cognitive components of pain perception in human beings and to investigate analogies in animals. The objective is to build a conceptual framework for the detection and alleviation of pain in animals.

WHAT IS PAIN?

Part of the difficulty of finding a universal definition of pain lies in the divergent empirical approaches to the problem adopted by physiologists, medical clinicians, behaviourists, philosophers and social scientists (Melzack 1968). For example, neuroscientists will often study the sensory phenomena associated with pain without reference to its motivational aspects. There can also be significant differences in definition. Pain may be described as a neurobiological process by neuroscientists but, in a philosophical context, it may encompass a much broader scope which includes 'emotional' pain and suffering.

The complexity of the pain phenomenon requires that any definition incorporates the neurobiological and psychological constructs and recognises the cognitive, emotional, social and cultural influences which determine pain perception. It must acknowledge that pain perception is associated with phenomena which can be established objectively and also with psychological, social or environmental factors which, through cognitive and emotional processes, can either of themselves be associated with pain or can modify the perception of pain associated with other phenomena. Thus, Melzack (1968) defined pain as a 'category of experiences, signifying a multitude of different unique events having different causes and characterized by different qualities varying along a number of sensory and affective dimensions'.

A model of pain based on an interactional approach which will recognise its multi-dimensional nature is important. Linear interactional models, proposed by Melzack (1973) and others, are useful in establishing causal links between the variables. However, the circular model proposed by Elton *et al.* (1983) may be more helpful, in that it allows for a better description of the complexity and interactive diversity of the various components.

The model for human pain proposed by Adams and Martin (1983) encompasses 'at least three components: nociception, the body's detection and signalling of noxious events; pain, the conscious perception or recognition of the nociceptive stimulus; and suffering, the affective, behavioural or emotional response to the pain'. The same three components underpin, but are not spelled out in, the working definition of pain in animals proposed by Zimmerman (1983) and modified by Kitchell (1987). 'Pain in animals is an aversive sensory and emotional experience (a perception), which elicits protective motor actions, results in learned avoidance, and may modify species-specific traits of behaviour, including social behaviour.'

The descriptions by Kitchell (1987) and by Adams and Martin (1983) take account of the three major experimental perspectives on pain which, when considered separately, have frustrated the formulation of an adequate global definition (Melzack 1968). These authors have dwelt on pain as either a sensory phenomenon or a drive-producer or an indicator of tissue pathology. All three aspects are important but are complementary, not mutually exclusive. Pain does classify with the sensations and is a sensory function of the nervous system. As such, it requires the activity of the brain, spinal cord, coaxial nerves and spinal nerves plus their ganglia and end-organs. Moreover, the higher adaptive function of pain and its value in protection or self-preservation of an individual requires the integrative activity of the brain. To clarify this, the brain takes signals from both the outside world and the internal environment, processes them, matches them against other factors and produces emotion, motivation and behaviour. Pain reception fits here as a sensory input.

PAIN, SUFFERING AND DISTRESS

In most situations, the terms 'pain' and 'suffering' are used inter-changeably, for example in citation indexes relevant to the biological and social sciences and to the humanities. Only in the *Philosophy Index* are they listed separately. Suffering is not mentioned in the *International Encyclopedia of the Social Sciences* (Sills 1968) and is discussed with 'pain' in the *Encyclopedia of Bioethics* (Reich 1978). In the latter, Shaffer (1978) notes that, although the concept of suffering is usually associated with pain, their relationship is complex and they are phenomenologically distinct. In the report of the AVMA Colloquium on Recognition and Alleviation of Animal Pain and Distress (AVMA 1987), it was concluded that 'suffering' is a much abused and colloquial term that is not even defined in most medical dictionaries. The AVMA (1987) defined suffering in animals as a 'highly unpleasant emotional response usually associated with pain and distress'. Similar difficulties in defining suffering were noted by Cassel (1982), who provided a definition of suffering in human beings as 'the state of severe distress associated with events that threaten the intactness of the person'. Thus, the concept of self-awareness is fundamental to Cassel's definition. Considerations of emotions, mental states and the mind are obviously more important to the definition and determination of suffering than to the concept of pain. Suffering is not a modality such as pain or fever. Further, it is possible to have pain without suffering and vice versa.

The association of suffering with mental states and self-awareness (consciousness) has led some, such as Bowsher (1980) and Craig (1984), to deny that animals can suffer—while acknowledging that they can perceive pain. In contrast, Zayan (1986) argues that feeling and emotion can be considered as elementary modes of self-awareness and, because they can be demonstrated in animals, are evidence that animals experience suffering.

Clearly, pain and suffering are not synonymous and, while it is possible to develop definitions of pain which include suffering, a widely accepted definition of suffering has yet to be formulated. For our purposes suffering will need to be considered both in the context of pain-related suffering and in the more general context which encompasses distress.

The model of human pain proposed by Adams and Martin (1983) has the operational advantage that the third and ultimate component, which determines the adaptive nature of pain and produces learned aversion, is identified as suffering. This third component embraces the concepts of emotion, motivation, affect, feeling-tone, memory, cognition and mental states as they relate to the phenomenon of pain.

The concept of distress will be introduced briefly as it relates to the previous discussion of suffering. 'Distress is a state in which the animal is unable to adapt to an altered environment or to altered internal stimuli' (AVMA 1987). Pain and distress may occur separately or in concert and either may lead to suffering. Fear and anxiety are common to all three states. Fear and anxiety may be the result of, or may modulate, the experience of pain or distress and are the predominant emotional components of suffering.

THE HUMAN MODEL OF PAIN

For convenience in analysing pain and pain-related suffering as it may occur through the animal kingdom, the three components of the model for human pain (that is, nociception, pain perception and affective–motivational responses) will be considered separately. It must be emphasised, however, that 'sensation, perception, and the various conscious and unconscious responses to a pain stimulus comprise an indivisible process' (Adams and Martin 1983). In addition, questions of 'awareness', 'consciousness' and 'mind' are intrinsic to the third component of the human model of pain and may be of variable significance to the second component. Hence they must be considered if the model is to be applied to animals.

In outlining the human model of pain, two kinds of pain with differing anatomical and physiological bases will be recognised (Nathan 1977). These are first or rapid pain (Bowsher 1983), which is acute and transient and usually does not outlast the inciting stimulus, and second or slow pain, which results from damage to tissue. The adaptive significance of the two kinds of pain may differ (Dennis and Melzack 1983). Aversifacient or rapid pain may serve as a response to the threat of tissue damage, by leading to rapid limb movement away from a noxious stimulus. The second kind of pain leads to tonic contraction of muscle, causing immobilisation and allowing healing and repair. This is seen in the 'favouring' of a limb, for instance. The dichotomy of pain here is physiological and not the division into 'sensory' pain and 'psychological' pain as described by Rachlin (1985).

NOCICEPTION: THE DETECTION OF NOXIOUS STIMULI

The first and absolutely essential event in the sequence which makes up the phenomenon of pain is the detection of tissue-damaging or potentially damaging stimuli by specialised neural end-organs, the nociceptors. In humans, the nociceptors in different tissues appear to have different adequate stimuli for the incitement of pain (Adams and Martin 1983). For instance, the stimuli for skin are clearly mechanical and injurious: pricking, cutting, burning and freezing. By contrast, similar stimuli do not operate for gut, where pain is related to inflamed mucosae, spasm of smooth muscle and tensions on peritoneal attachments. For muscle, the adequate stimulus appears to be ischaemia. Other tissues have other adequate stimuli.

In mammals, two types of cutaneous nociceptors and their afferent nerves have been described. Small myelinated nerve fibres and A-delta nociceptors signal first or rapid pain, whereas small unmyelinated nerve fibres and C-polymodal nociceptors signal second or slow pain (Kruger and Rodin 1983). Myelinated fibres transmit signals faster than unmyelinated fibres. A-delta nociceptors have a threshold for mechanical stimulation which is a thousandfold higher than for low-threshold mechano-receptors which are concerned with touch. They are also very responsive to heat. The C-polymodal receptors are triggered by a wide range of substances. Chemical mediators including histamine, serotonin and the potent pain-inducer bradykinin have been implicated, as have intracellular substances such as potassium ions, hydrogen ions, acetylcholine, adenosine phosphates and lysosomal enzymes (Spector 1980, Adams and Martin 1983, Kitchell 1987). Another group, the prostaglandins, acts by lowering the threshold of responses to other agents, and by prolonging the sensation.

In humans, pain due to stimulation of A-delta nociceptors is said to be more severe than that signalled by the C-fibres when a single electric shock is given. However, C-fibre pain appears to be additive and particularly aversive, and leads to severe suffering (Burgess 1978).

Although knowledge of nociception has been obtained in the main with cats and primates, nociceptors exist in all mammals (Wall and Melzack 1984) and birds (Breward and Gentle 1985). There may, however, be differences in the neurotransmitters involved (Pierau *et al.* 1985). How can nociception be identified in other taxa? The primary criterion may be the presence of clearly aversive behaviours in response to those stimuli associated with nociception in mammals. Avoidance behaviour stimulated by touch, sight, hearing and smell must be discriminated. In other words, the adequate stimulus must specify nociception. As illustration, the withdrawal reflex stimulated by squeezing the claw of pithed toads is evidence that nociceptive pathways exist in this amphibian. The responses of *Octopus* and *Sepia* species to electric shock (Boycott 1965, Messenger 1977) indicate that nociception occurs in these cephalopods. Embryological and anatomical evidence in fish, amphibians and reptiles indicates that in these species, neural reflexes are modulated by internuncial fibres such that behaviour is not simply a reflex but becomes 'variable, unique and creative' (Whiting 1955).

Secondary criteria for the possibility of nociception in a given group of animals may be the existence of structures in the peripheral nervous system similar to those concerned with nociception in mammals. If the inflammatory mediators—histamine, serotonin or bradykinin—can be found and assigned a role in inflammation in a given species, these substances may also trigger polymodal nociceptors when tissue is damaged. If they cannot be found, other substances may act analogously; or nociception may only be possible through high-threshold mechanical or heat receptors, or through polymodal chemo-receptors with a limited range of adequate stimuli.

Knowledge of nociception throughout the phylogenetic tree appears to be scarce and may only exist in obscure form. The question of nociception as opposed to avoidance mediated by other and innocuous stimuli has not, in most cases, been directly addressed. Without nociception, pain and pain-induced suffering based on the human model of pain perception cannot exist. Evidence supports the existence of nociception in all vertebrates, and possibly in some invertebrates such as cephalopods.

PAIN: THE PERCEPTION OF NOXIOUS STIMULI

Several aspects of this second component of the model for human pain, the perception of nociceptive stimuli, may be important when the potential for pain in animals is being considered. These include: (1) the specific neural circuitry involved, especially when animals such as mammals with a similar neuroanatomy are being considered; (2) the presence of neurons and neural pathways with receptors and sensitivity to analgesics and other pain-modifying compounds; and (3) the presence of mechanisms which act by inhibition rather than excitation. Other areas which should be explored include the idea that pain is a perception, an elaboration of sensation, which may result from the coexistence of 'fast' and 'slow' pain and the subqualities which may occur within each of these modalities; and the idea that the perception of pain and the accompanying aversive feeling-tones may be discriminated and dissociated at a physiological level.

Spinal aspects of pain perception

According to the human model, nociceptors transmit their signals to the central nervous system where the phenomenon of pain is synthesised from activity in the spinal cord, brainstem, thalamus and cerebral cortex (Adams and Martin 1983, Casey and Morrow 1983, Bowsher 1984, Iggo 1984, Willis and Chung 1987). The perception of pain then gives rise to the possibility of the motivational–affective response of pain, the third component of the model.

The neural circuitry involved in pain can be traced from where nociceptive signals enter the dorsal horn of the spinal cord and relay both within the same segment and to other segments. This relay activates local spinal reflexes such as the flexion reflex and crossed extensor reflex, which are concerned with escape from noxious stimuli. Other connections in the spinal cord are made with ascending tracts which then transmit nociceptive signals to the brain. As described in the gate control theory of pain (Melzack and Wall 1965), cells in the dorsal horn of the spinal cord which are 'excited by these injury signals are also facilitated or inhibited by other peripheral nerve fibres which carry information about innocuous events' (Wall 1978). In addition, certain centres in the forebrain and brainstem—for example, the periaqueductal grey matter and the nucleus raphe magnus—can signal through descending tracts in the spinal cord and modulate the excitability of those cells which transmit information about injury (reviewed by Fields and Basbaum 1978). The existence of mechanisms which act by inhibition rather than excitation is crucial if pain in animals is to be gauged by the human model of the phenomenon. Without inhibition/excitation switches, responses to noxious or potentially noxious stimuli are bound to be fixed, stereotyped and lacking the potential for adaptive modification or elaboration. If this is so, simple mechanisms can account for them and the complex and highly integrated central nervous component which allows for pain perception in human beings may not be invoked.

In humans, two ascending tracts in the spinal cord are activated by nociceptive stimuli (Adams and Martin 1983, Bowsher 1983, 1984). The neospinothalamic pathway is believed to be concerned with the sensory–discriminatory aspect of pain, in that it transmits information on its location and intensity. This pathway relays through the medulla, pons and midbrain to the thalamus and thence to the primary somatosensory cortex. The other pathway, the palaeospinothalamic tract, including the spinoreticular and spinomesencephalic, is believed to mediate the arousal, affective and emotional aspects of pain and second or slow pain. This tract connects through the reticular formation in the brainstem to the hypothalamus and thalamus, with connections to the limbic system and thence to the secondary somatosensory cortex.

In those species which have been studied—mainly the monkey, cat, rat and pig—there are major differences in the relative importance of the various spinothalamic tracts and in the details of their origin and destination (Melzack and Dennis 1980, Ralston 1984, Willis 1984, Willis and Chung 1987). The palaeospinothalamic tract is phylogenetically old but has maintained its functional significance throughout mammalian evolution. The neospinothalamic tract develops greater significance higher up the evolutionary tree. This development parallels that of the somatosensory thalamus and cortex as sensory discrimination becomes an important component of pain perception. The neospinothalamic tract is comparatively less influential in lower mammals and does not achieve prominence until the evolution of primates (Melzack and Dennis 1980).

The spinocervicothalamic tract is of relatively minor importance in primates and does not exist in a proportion of human beings (Truex *et al.* 1970). It is well-developed in some lower mammals, especially the cat, in which it has been extensively studied (Jones *et al.* 1985). The spinocervicothalamic tract and the postsynaptic dorsal column pathways are both alternative, rapidly conducting pathways which have been demonstrated in sub-primate species, and in those such as the cat are probably major nociceptive ascending tracts (Willis and Chung 1987).

Studies by Ebbesson (1969) demonstrated spinothalamic elements in fish, amphibians and reptiles, consisting of small-diameter, unmyelinated fibres which project to the reticular area in the brainstem. Melzack and Dennis (1980) concluded that the existence of these reticular neurons and the associated limbic structures are evidence of the motivational–affective system in these species. The neospinothalamic system is a later evolutionary development.

Thus, while there has been comparatively little change in the palaeospinothalamic tract, further up the phylogenetic tree there are important changes in the development and significance of the spinocervico- and neospinothalamic tracts. An intact spinothalamic tract is essential for pain perception in human beings (Price *et al.* 1978). All vertebrates possess neural connections between peripheral nociceptors and central nervous structures—a prerequisite for pain perception.

Supraspinal mechanisms

Nociceptive stimuli transmitted by the spinothalamic tracts are relayed either through the nuclei of the thalamus to the sensory–cerebral cortex, or through the reticular formation to the thalamus and hypothalamus and then to the limbic system. Once again, inter-species differences are significant, especially as they relate to sites of termination of various spinothalamic tracts (Berkley 1980, Guilbaud *et al.* 1984). Further, there are important ultrastructural differences, particularly in the lateral thalamus (McAllister and Wells 1981, Ralston 1982).

The thalamus and the primary sensory cortex are the supraspinal structures implicated in pain perception. The reticular formation, hypothalamus, limbic system and secondary somatosensory cortex modulate pain perception. The involvement of these centres has been demonstrated by specific stimulation and/or inhibition studies, and by observation of pain-related behaviour in both human beings and animals (see Kitchell and Johnson 1985). Reciprocal relationships can be demonstrated between particular regions of the primary somatosensory cortex and peripheral noxious stimuli (Ralston 1984). Recent evidence indicates that the role of the thalamus is complex and intricate and is not that of a simple relay system. The thalamic nuclei are so organised that a wide variety of synaptic interchanges occur between spinothalamic projections, the brainstem and neurons of the diencephalon, and the cerebral cortex (Ralston 1984).

There is a general acceptance, but little direct evidence, that the perception of pain involves activity in the cerebral cortex. Present methodology is the major limitation to studies in this area, but recent evidence of a correlation between cerebral-evoked potentials and the subjective experience of pain in humans is promising (Chudler and Dong 1983). Similar correlations in animals, but with pain experience judged by behavioural measurements, would provide evidence that the perception of pain in people and other animals is similar (Molony 1986). Other approaches to cerebral cortical function, including electroencephalography and the measurement of cerebral metabolism, may assist in such studies (Molony 1986).

There are anatomical differences between species in the cerebral cortex. These may have important implications for pain perception and the motivational–affective modulation of pain. All mammals have substantial areas of primary somatosensory cortex. With increasing complexity of neural organisation there is significant variation in the secondary cortex. The degree of cortical folding augments the relative amount of secondary cortex. Accordingly, animals with a gyrencephalic or furrowed cortex—for instance higher primates—have more secondary cortex relative to total brain volume as compared with animals with a lissencephalic or smooth cortex—such as cattle and sheep. Primitive gyrencephaly is seen in cats, dogs and horses (Bowsher 1980). The development of the prefrontal cortex is associated with a significant increase in secondary cortical area; this is a distinctive development of the higher primates, and especially of human beings.

It cannot be claimed that this greater area of secondary cortex, particularly in the prefrontal region, directly underlies the characteristics of intelligence and consciousness in humans. Nevertheless, there are associations and these suggest links with the motivational–affective and cognitive aspects of pain. Furthermore, the prefrontal area is the part of the secondary cortex closest to the hypothalamus which, in turn, is the supraspinal centre implicated in transmission from the palaeo-spinothalamic tract to the limbic system, and in the motivational–affective aspects of pain perception. There is functional but not anatomical evidence of connections between the hypothalamus and prefrontal cortex (Bowsher 1980).

As noted by Melzack and Dennis (1980), one of the most striking features of pain expression is the increasing influence of cognitive facilities higher up the phylogenetic scale. Cognitive information based on past experience and probable assessment of outcomes can modulate activity and responses in both the discriminative and motivational systems. In this connection, an indication of the significance of the prefrontal cortex in pain perception in human beings is provided by the reports of patients who have undergone frontal lobotomy. They experience significant changes in the motivational–affective aspects of pain. When a noxious stimulus is applied, they feel pain but are not bothered by it—it is not aversive (Foltz and White 1962).

Neural analogy permits the possibility of pain perception in all vertebrates (Melzack and Dennis 1980) and the pain detection threshold is similar across the vertebrate species, including human beings (Kitchell and Johnson 1985). Nevertheless, the development of rapid sensory–discriminatory ability and the greater influence of cognition in the modulation of pain perception as seen in some mammals, particularly primates and especially human beings, raises the prospect of quantitative or even qualitative differences in pain experience.

In human beings the palaeospinothalamic system appears only to play a role in pain perception in patients with deafferentation pain (phantom-limb syndrome). Stimulation of the medial thalamus produces pain, although in normal patients this would not occur (Albe-Fessard *et al.* 1984). The neurological basis of this deafferent syndrome is unclear. It may have implications for pain perception in those species where the palaeospinothalamic is the only or major spinothalamic tract.

Endogenous pain control

Endogenous pain-control mechanisms are important in modulation of pain perception and adaptation to painful stimuli. Evidence for such mechanisms is based on two sets of observations (Basbaum and Fields 1984). First, there is considerable evidence of descending pain-control circuits which are extremely complex, originate in the mid- and hind-brain areas and project to central laminae in the dorsal horns of the spinal cord. The periaqueductal grey matter and the nucleus raphe magnus of the caudal brainstem are two major components of this system. There are a number of complex and highly integrated inhibitory circuits involving the supraspinal centres and local feedback loops in the spinal cord. These systems may modify the pain response and may alter pain perception when they are associated with hypo- or hyperalgesic states.

However, it is important to realise that descending inhibition is not equivalent to stimulation-induced analgesia. These descending inhibitory mechanisms have been demonstrated in a number of mammalian species but have been most extensively studied in the cat, monkey and rat (Carstens 1987). The neuropeptides implicated in these systems are predominantly serotonergic, although endogenous opiates, vasopressin, neurotensin and substance P may be involved. Noradrenergic involvement in the control and feedback of this system has been suggested, but evidence is not conclusive (Fields and Basbaum 1984).

The second set of observations relating to endogenous pain control involves the opiate system. There are at least three families of endogenous opioid peptides—enkephalins, dynorphins and β-endorphins. While enkephalin and dynorphin are extensively distributed throughout the body, β-endorphin-containing neurons are restricted to the central nervous system (Bloom *et al.* 1978). Also, there are three populations of opiate receptors, μ, $\varkappa$ and θ. Analgesia is mediated through μ-receptors (Audigier *et al.* 1980, Kosterlitz *et al.* 1980). They are involved in decreasing nociceptive flexor reflexes, and show strong naloxone antagonism. $\varkappa$-Receptors are associated with weak analgesic activity and are mainly depressant, while θ-receptors have mainly excitatory properties. β-Endorphin has a high affinity for all three receptors, although there are differences in receptor affinities for the other families of opioid peptides (Wüster *et al.* 1981). Finally there are a number of neuropeptides which act as endogenous antagonists to the opiate system (Bhargava 1982, Watkins *et al.* 1986).

Despite extensive studies, the physiological role of endogenous opiates is not clearly defined. Studies show that endogenous pain suppression occurs when the neural circuitry is activated, and that endogenous opiates play a central role in the activation and modulation of this system. The interrelationship of neuronal circuitry and opioid peptides is complex, and multi-directional models have been proposed to explain the intricacy of the modulating effects observed (for example, Gillman and Lichtigfeld 1985). With the multiplicity of peptides, receptors and neural components involved, significant inter-species differences are likely but little comparative data are available. Moreover, it is not known how different components of the system would qualitatively influence pain perception. Nevertheless, endogenous pain-modulating mechanisms probably exist in all mammals (Pert *et al.* 1974). There is also evidence that endogenous opiates occur in invertebrates such as the earthworm (Alumets *et al.* 1979). Their presence alone does not prove a capacity to perceive pain. It does, however, suggest an early involvement of endogenous opiates in the evolution of pain perception mechanisms.

Stress-induced analgesia

While endogenous pain-suppression mechanisms can be demonstrated in the laboratory, their role in pain perception under normal circumstances is not clear. A variety of painful or stressful events induce analgesia (Amir *et al.* 1980, Chance 1980). Amit and Galina (1986) proposed that this stress-induced analgesia is an adaptive response in that it permits motor activity to alleviate the noxious stimulus. The animal can react to certain situations as if pain perception were below the tolerance threshold. An interesting example of this is the activation of the system during pregnancy and the associated elevation of pain threshold during parturition (Rust *et al.* 1984).

Depending on the type, intensity and duration of the stimulus, two types of stress-induced analgesia are seen—one is opioid dependent, the other non-opioid (Terman *et al.* 1984). Observations suggest that they operate as separate endogenous analgesic systems that mutually inhibit each other (Weinstein *et al.* 1988), although there is some commonality in the neural circuitry involved (Amit and Galina 1986). For our considerations it is important that both systems can be activated by nociceptive stimulation and by various psychological stresses, such as exposure to a novel environment. Further, the response can be induced by anticipation of pain (Jakoubek 1984). As reviewed by Amit and Galina (1986), the expression of stress-induced analgesia will depend upon and can be modified by a variety of environmental conditions. Insofar as it is a critical component of the behavioural response to a threatening stimulus, it must involve supraspinal mechanisms and is therefore an important component of the motivational–affective response to pain.

There are few comparative data on stress-induced analgesia between the species. Recent studies by Nolan *et al.* (1987) suggest that the non-opioid system predominates in the sheep. Information like this is most relevant to the choice of analgesic agents. Knowledge of inter-species differences in the endogenous pain-control system will allow the selection of more effective means of alleviation. However, it is not known how such differences could influence pain perception.

MOTIVATIONAL–AFFECTIVE ASPECTS OF PAIN

For an animal to perceive a noxious stimulus such as pain, the sensory input must be coupled with a neural system which has the capacity to interpret that stimulus. The evaluation of the stimulus will, at the least, be based on previous experience. Hence, plasticity (i.e. neural memory) of the central/interpretative neural mechanism is important. In primates, perception mainly involves the secondary and tertiary sensory areas of the cortex (Zayan 1986). For animals to perceive pain they must possess an analogue of the secondary cortex at least. The role of these areas of the cerebral cortex in motivational–affective functions suggests *per se* their significance in the perception of pain. Growth and development of these areas, and consequent cognitive function in higher mammals (particularly primates), also suggest that motivational–affective components may assume greater complexity and importance in the pain experience of these animals.

The evaluation of the noxious stimulus becomes a complex analysis of multimodal information encompassing past experiences and evaluation of possible outcomes. The perception of pain requires some cognitive modality, and hence there must be cerebral cortical involvement in both discriminative and motivational systems that influences reactions based on cognitive evaluation. The minimal level of cognitive development for pain perception remains speculative. Zayan (1986) suggests that only the most rudimentary level is required. On this basis, pain perception is recognised in all vertebrate species. The capacity to anticipate the consequences of events is suggested as the significant difference in the motivational–affective component of human and animal pain (Craig 1984). It is interesting that the spinothalamic tract involved in the motivational–affective component of pain is phylogenetically of long standing. This raises the question of the possible importance of this modality in the evolutionary development of pain.

Emotion and anxiety

Emotion is an important ingredient in the motivational aspect of pain perception. It is widely recognised that the perception of pain can be attenuated or accentuated by emotion (Craig 1984). The existence of emotions in animals is inferred by neural analogy (Crosby and Showers 1969, Parent 1979) and there is now widespread agreement that animals experience emotion (Plutchik 1980). Panksepp (1982) proposes a general psychobiological theory of emotions for human beings and mammals. He defines emotion as a 'generalised arousal state from which individual emotions evolve via a process of social learning'. The main thesis of his paper is that there are four primary limbic 'command' circuits—fear, expectancy, panic and rage—which receive a variety of external stimuli, and that various interactions of these command circuits result in a range of specific behavioural responses. He proposes that subjective human experience can be used to analyse and categorise the role of these neuronal circuits in the expression of emotions.

Stephens (1988) has reviewed the various experimental approaches to analysis of emotional behaviour in farm animals, in particular the behavioural evidence of emotional motivation and the concomitant evidence of physiological and endocrinological change. Such an approach provides the basis for demonstration and evaluation of the motivational–affective responses in animals.

Anxiety is a significant emotional modulator of pain perception. Increased anxiety potentiates pain (Sternbach 1974, Schumacher and Velden 1984), whereas a reduction in anxiety will have an ameliorating effect on pain perception (Craig 1984). Definition of a neurochemical basis for anxiety with specific benzodiazepine receptors in the central nervous system will assist in the demonstration and evaluation of anxiety states in human beings and animals (Gee *et al.* 1984, Richards and Mohler 1984). Nielsen *et al.* (1978) demonstrated the presence of benzodiazepine receptors in a range of vertebrate species including fish, but failed to detect significant levels in invertebrates. Studies in monkeys (Ninan *et al.* 1982) and rats (Vogel *et al.* 1971) demonstrate that benzodiazepine receptors are involved in both the affective and physiological manifestations of anxiety in these species, but species variation is seen (Crawley *et al.* 1984).

Benzodiazepine receptors have been shown to be involved in the regulation of secretion of pituitary β-endorphin (Maiewski *et al.* 1985) and thyrotropin (Boyadjieva and Ovtcharov 1987), and thus may modulate the endogenous pain-control systems.

Because emotion is the affective response to an evaluation and interpretation of sensory inputs from the internal and external environment, evidence of its existence supports the possibility of pain-related perception. Fear is a primitive state of mind (Cassano 1983), and becomes the feeling of anxiety through a 'conscious' evaluation involving retrospection and imagination (Jaynes 1982). Anxiety is directed towards the future (Lewis 1980) and is based on possibility. Hence, evidence of anxiety in animals suggests the ability, however rudimentary, to evaluate possible outcomes. However, the level of cognitive development is important in anxiety states, particularly as it relates to abstract thinking, fantasy and imagination. These abilities are most highly developed in human beings and there is considerable debate about their existence in other animals.

Cognitive development and pain perception—awareness, consciousness and mental states

Cognitive appraisal in human beings is a major determinant and modulator of pain perception (Craig 1984). As defined by Doré and Dumas (1987) 'cognitive development and its product, knowledge, are the result of a continuous interaction between the subject and its environment'. It is a process whereby learned information is transferred, thus providing a more general and flexible representation. Intelligence is the biological and psychological adaptation that makes flexible programming of behaviour possible (Piaget 1967).

The model for cognitive development is a human one. A schema developed by Piaget (1967) is used to determine and evaluate the origins, nature and ontogeny of human knowledge. In this model cognitive development is divided into four major phases. The first is the sensorimotor period, from birth to about 2 years, during which a child lacks mental agility and can only deal with real actions in a narrow range of space and time. The pre-operational period (2 to 7 or 8 years) represents a phase of development preparatory to the concrete operational period (7 or 8 to 11 or 12 years), during which a child can internalise events and understand more complex relationships between them. The formal operational period (11 or 12 years and older) is the adult mode, characterised by the ability to reason logically, starting from premises and drawing conclusions. Doré and Dumas (1987) have written a review of the application of the Piagetian model to the study of animal cognition. There is evidence of a sensorimotor phase in a number of species, mainly primates but also cats, dogs and wolves. Chimpanzees, gorillas and human infants go through comparable development during early life and there are signs of rudimentary cognition in other species. However, despite these similarities there are important structural differences. Other Piagetian phases have been identified in animals, mostly in primates. Based on evidence for sensorimotor intelligence, object permanence, causality and imitation, it is possible that the great apes (chimpanzees, gorillas and orang-utans) not only reach all stages of the sensorimotor period but also have a rudimentary form of pre-operational intelligence. It is interesting to note the similarities in these conclusions to the scheme proposed by Romanes (1888). However, Doré and Dumas (1987) emphasise that there are significant methodological problems, and that, to evaluate animal cognition and intelligence truly, more attention should be given to the use of species-specific characteristics.

Consideration of cognitive development encompasses such concepts as awareness, consciousness and mental states. There is considerable reluctance to attribute any of these qualities to animals. None are mentioned in the most recent edition of the *Oxford Companion on Animal Behaviour* (McFarland 1985), and in a major symposium on animal cognition (Roitblat *et al.* 1984) a sharp distinction was made between animal and human cognition. Indeed, even a possible discontinuity in evolutionary development was proposed (Terrace 1984). Two recent texts—*Mental Models* (Johnson-Laird 1983) and *The Psychobiology of Consciousness* (Davidson and Davidson 1980)—discuss consciousness and mental states as exclusively human characteristics. Davidson (1980) proposes that 'one function of consciousness is to transfer and restructure information'. There is suggestion of a rudimentary capacity of this function in animals which exhibit expectancy. It is also a component of anxiety states. Perhaps, once again, the appropriate human analogy is found in children. Izard (1980) suggests that although infants possess the 'core of consciousness', it is not until the second half of the first year of life, when affective experience develops some independence of direct sensory contact, that a level of consciousness is achieved.

The existence of consciousness and of mental states in animals is a matter of considerable contention (Griffin 1984, Burghardt 1985, Gallup 1985). This debate is confounded by conflicting definitions of awareness, consciousness and mind. Gallup proposes a working definition of 'consciousness' as being aware of your own existence, and of 'mind' as being the ability to monitor your own mental states and the corresponding capacity to use your experience to infer the experience of others. The studies of Gallup (1985), Premack (1984), Savage-Rumbaugh *et al.* (1983) and Ristau (1983) indicate rudimentary forms of self-awareness (consciousness) and mental states in the great apes. Reciprocal altruism has been observed in chimpanzees (de Waal 1982) and rhesus monkeys (Masserman *et al.* 1964). Thus mental continuity between human beings and other animals must be considered (Griffin 1981, Ristau 1983).

Language is often cited as the primary evidence for the existence of consciousness and mental states in human beings but not in other animals. Recently, questions have arisen about the existence of rudimentary language in primates (Froehlich 1984).

There are significant differences in cognitive development between human beings and other animals. Nevertheless, the emergence of consciousness and mental states and the intellectual capacity for introspection and abstract thought similar to that in human beings can be demonstrated in a rudimentary form in some animals. This is most obvious in the great apes, where the level of cognitive development may be comparable to that seen in very young children. There is no evidence of animals achieving cognitive development comparable to that associated with introspection and abstract thought in humans (Piagetian concrete operational). These mental capacities are important to the complexity and significance of the motivational–affective modulation of pain perception in human beings.

There have been few studies of cognitive development and pain perception in children. Beales *et al.* (1983) showed significant differences in pain perception in young (6 to 11 years) compared to older children (12 to 17 years) and suggested that this was related to the level of cognitive development. They reported that the younger children 'felt' the pain but it did not matter so much to them. However, the existence of specific behavioural changes after circumcision in the neonate indicates that memory is an important sequel to pain perception even at this age (Anand and Hickey 1987). Possibly the recent interest in pain in human neonates and children will provide data that will assist in understanding the significance of cognitive development and modulation of pain perception. Such data also would be relevant to consideration of those modalities in pain perception in other animals.

There is evidence of motivational–affective modulation of pain perception in animals. Jakoubek (1984) reported a reduction in pain threshold in rats which had been habituated to handling, and this was augmented by treatment with anti-anxiety agents. Similar results were obtained by Jørum (1988) who also found a correlation between the emotional behaviour of the rats and the development of hypo- or hyperalgesia. Similarly, Stanley (1987) reported a decreased dose requirement for an anaesthetic agent in calves which had been handled. It is possible that these modulations are mediated through alteration in the anxiety state. Of relevance is the work of Corda *et al.* (1980) who demonstrated significant differences in cerebral nucleotides involved in GABAergic transmission in rats habituated to handling.

The recognition of a motivational–affective component of pain perception and the contribution of anxiety has important implications for the detection and alleviation of pain in animals. The possibility of modulating pain perception through the alleviation of anxiety, using both pharmacological and non-pharmacological means, warrants attention and further study (Wolfle 1987).

Behavioural evidence of pain perception

As noted by Zayan (1986), neural analogies are a necessary but insufficient condition for inferring that animals feel pain. He contests that additional analogies, such as behavioural indicators, are needed.

Pain is both a determinant and a modifier of behaviour. Pain-related behaviour not only is indicative of pain perception but also is the expression of the motivational–affective component. As noted by Craig (1984), the affective component of pain is closely linked to motivational properties. A major difference in the affective–motivational expression of pain in human beings and animals is language (Ehlich 1985). In both humans and animals, behavioural responses to pain are either for the purpose of avoiding the stimulus or for seeking assistance (by vocalisation), or both. In chronic pain, a different behavioural repertoire is developed. To define an observed behaviour as pain-related it is necessary to show that it is not a simple reflex. Such evidence can be derived from evidence of learning and adaptation or from the modification of the behavioural response by pharmacological inhibition or stimulation. Here, the human analogy is helpful. Wall (1979) proposed that traumatic injury in humans and other mammals is followed by three phases of behaviour which have the same general form.

First is the immediate phase where activities for avoidance and defence take priority and where pain is not always evidenced. The emergency, alarm or fight reaction—the co-ordinated result of the activated sympathetic nervous system and secretions released from the adrenal medulla—is involved (Tietz and Hall 1977).

Second is an acute phase 'a transition between coping with the cause of the injury and preparing for recovery', characterised by a 'combination of tissue damage, pain and anxiety' (Wall 1979). This phase is not so well described for animals.

Third is the sequence of pain-induced behaviour as a chronic phase characterised by quiet inactivity. 'Quiet inactivity might be the optimal tactic to encourage cure and recovery of damaged tissue' (Wall 1979). This idea has been used by Bolles and Fanselow (1980) in their perceptual–defensive–recuperative model of fear and pain.

All vertebrates demonstrate a capacity for learning and discrimination, and behavioural evidence of pain (Zimmerman 1986) which has been the basis for successful detection and alleviation of pain in these species (Vierck and Cooper 1984, Wright *et al.* 1985, Morton and Griffiths 1985, Sanford *et al.* 1986). Other species such as cephalopods have also demonstrated the capacity for learning and discrimination (Young 1961). The possibility of pain perception in invertebrates has been reviewed recently by Fiorito (1986). Opiate-like activity is seen in a number of invertebrates and opiate peptides modulate the responses to stimuli that are both nociceptive and elicit self-preserving behaviour.

The reliability of behavioural evidence of pain perception is improved if a number of indicators are measured (Zayan 1986). The repertoire of behavioural responses to pain is species-specific and will vary between individuals and within individuals depending on a range of environmental and social conditions. For example, the response of rats to painful stimuli is influenced by social factors and by contact with conspecifics (Williams and Eichelman 1971).

Finally, it is interesting to speculate on the relationship between the function of endorphins and pain-related behaviour. Endorphins play a significant role in animal learning and behaviour (Riley *et al.* 1980):

they inhibit distress vocalisation (Herman and Panksepp 1981) and modulate food and water intake (Vaswani *et al*. 1983). Vocalisation and food and water consumption are important and reliable indicators of pain in animals (Morton and Griffiths 1985).

HUMAN PAIN AND ANIMAL PAIN

Based on the human model of pain (Adams and Martin 1983), pain perception can be implied in all vertebrate species on the basis of neural analogy and evidence of pain-associated behaviour. On this basis it is not possible to impute pain perception to invertebrates, but the evidence of pain-associated behaviour in some invertebrate species indicates that the possibility of pain perception in these species must be considered. There is evidence of all three components—nociception, perception and motivational–affective—in mammals. The capacity for motivational–affective modulation of pain in lower vertebrates such as fish, amphibians and reptiles is not known.

Modulation of pain perception by emotional and cognitive modalities can be demonstrated in mammals. In these species anxiety can potentiate pain perception. The level of cognitive development, particularly in the great apes, indicates that cognitive processes are likely to have a greater influence on pain perception in these species.

How and if inter-species differences in neural circuitry influence either the qualitative or quantitative aspects of pain perception is not known, but it is likely that such differences are significant to the choice of agents to alleviate pain.

The major difference between human beings and other animals lies in the cognitive modalities of pain perception. Cognition is a major determinant of pain perception in human beings. Few animal species have been shown to have more than the most rudimentary cognitive development comparable to the level of young children.

This difference in cognition relates not to the perception of pain *per se* but only to the affective–motivational modulation of that perception. This greater capacity for cognitive modulation means that human beings have a greater potential either to attenuate or to limit pain.

DO ANIMALS SUFFER?

> The question is not,
> Can they Reason? nor, Can they talk?
> but, Can They Suffer?

This assertion by Bentham (1789) in *An Introduction to the Principles of*

Morals and Legislation, has been the basis for the development of utilitarian consideration of animal interests in recent times.

As stated in the introduction to this discussion, suffering is a poorly defined and much misused term. Suffering in the human context, as discussed by Cassel (1982), is reported when people perceive a threat to their integrity as persons. Such circumstances occur 'when they feel out of control, when the pain is overwhelming, when the source of pain is unknown, when the meaning of the pain is dire or when the pain is chronic'. Human suffering can be associated with either physical or psychological events, or both, and incorporates conscious processes (i.e. self-awareness) and highly complex cognitive capacities involved in the anticipation of the consequences of events—introspection, abstract thought and fantasy. There is neither a morphological nor a physiological determinant of suffering in humans.

On the basis of these complex cognitive processes and the role of self-awareness or consciousness, it has been claimed that suffering is a uniquely human experience (Bowsher 1980, Craig 1984). However, on the basis of comparative cognitive studies, the rudiments of human suffering must be acknowledged at least in the great apes.

Nevertheless, there is argument by Dawkins (1980, 1986), Fraser (1984/85), Zimmerman (1986) and Zayan (1986) that animals do suffer. Zimmerman (1986) contests that awareness is sufficient evidence for animals to be able to suffer. The attribution of suffering to animals is important, in that it acknowledges that animals can experience emotional distress not only caused by physical pain but also in the absence of physical disease. However, whether this should be considered as suffering in the human context should be examined. The subject of suffering warrants closer scrutiny in both people and animals.

IMPLICATIONS

This review of pain and suffering in other animals raises more questions than it answers. Possibly the greatest difficulty that we have in furthering our understanding of pain and suffering in other animals is the limitation of the human model as applied to other species. Further studies will need to explore the significance of the development of cognitive processes, and the human analogy may well be inappropriate.

Evidence of pain perception in animals nevertheless provides the basis for devolving responsibility onto humans to alleviate or minimise pain. The complexity of modalities which may influence pain perception, and the differences that may occur in individual expression, mean that pain must be evaluated on an individual animal basis. Further, evidence for involvement of the motivational–affective component in the modulation of the pain experience indicates that greater emphasis should be given to the alleviation of pain in animals by influencing these components.

REFERENCES

Adams, R. D. and Martin, J. B. (1983). Pain. In Reterdorf, R. G., Adams, R. D., Braunwold, E., Isselbacher, K. J., Martin, J. B. and Wilson, J. D. (eds), *Harrison's Principles of Internal Medicine*, 10th edn, McGraw-Hill, New York, pp. 7–15

Albe-Fessard, D., Condés-Lara, M., Sanderson, P. and Levante, A. (1984). Tentative explanation of the special role played by the areas of paleospino-thalamic projection in patients with deafferentation pain syndromes, *Adv. Pain Res. Ther.*, **6**, 167–81

Alumets, J., Håkanson, R., Sundler, F. and Thorell, J. (1979). Neuronal localisation of immunoreactive enkephalin and β-endorphin in the earthworm, *Nature*, **279**, 805–6

Amir, S., Brown, Z. W. and Amit, Z. (1980). The role of endorphins in stress: evidence and speculations, *Neurosci. Biobehav. Rev.*, **4**, 77–86

Amit, Z. and Galina, Z. H. (1986). Stress-induced analgesia: adaptive pain suppression, *Physiol. Ref.*, **66**, 1091–120

Anand, K. J. S. and Hickey, P. R. (1987). Pain and its effects in the human neonate and fetus, *New. Engl. J. Med.*, **317**, 1321–9

Audigier, Y., Mazarguil, H., Gout, R. and Cros, J. (1980). Structure–activity relationships of enkephalin analogs at opiate and enkephalin receptors. Correlation with analgesia, *Eur. J. Pharmacol.*, **63**, 35–46

AVMA (American Veterinary Medical Association) (1987). Colloquium on Recognition and Alleviation of Animal Pain and Distress, *J. Am. Vet. Med.*, **191**, 1186–296

Bankowski, Z. and Howard-Jones, N. (eds) (1984). *Biomedical Research Involving Animals*, CIOMS, Geneva

Basbaum, A. I. and Fields, H. L. (1984). Endogenous pain control systems: brainstem spinal pathways and endorphin circuitry, *Annu. Rev. Neurosci.*, **7**, 309–38

Bateson, P. G. (1986). When to experiment on animals, *New Sci.*, **109**, 30–2

Beales, J. G., Kean, J. H. and Holt, P. J. L. (1983). The child's perception of the disease and the experience of pain in juvenile chronic arthritis, *J. Rheum.*, **10**, 61–5

Bentham, J. (1789). In Harrison, W. (ed.) (1967). *An Introduction to the Principles of Morals and Legislation*, Blackwell, Oxford, p. 412

Berkley, K. (1980). Spatial relationship between the terminations of somatic sensory and motor pathways in the rostral brainstem of cats and monkeys. I. Ascending somatic sensory inputs to lateral diencephalon, *J. Comp. Neurol.*, **193**, 283–317

Barghava, H. N. (1982). Effect of peptides on the development of tolerance to buprenorphine, a mixed opiate agonist–antagonist analgesic, *Eur. J. Pharmacol.*, **79**, 117–23

Bloom, F., Battenberg, E., Rossier, J., Ling, N. and Guillemin, R. (1978). Neurons containing β-endorphin in rat brain exist separately from those containing enkephalin: immunohistochemical studies, *Proc. Natl. Acad. Sci.*, **75**, 1591–5

Bolles, R. C. and Fanselow, M. S. (1980). A perceptual defensive recuperative model of fear and pain, *Behav. Brain Res.*, **3**, 291–323

Bowsher, D. (1980). Pain sensations and pain reaction. In Wood-Gush, D. G. M., Dawkins, M. and Ewbank. R. (eds), *Self-Awareness in Domesticated Animals*, Universities Federation for Animal Welfare, Hertfordshire, pp. 22–8

Bowsher, D. R. (1983). Pain sensations and pain reactions. In Papers prepared by members of the Royal Society's Ethical Working Party on Animal Experiments, The Royal Society of London, pp. 24–8

Bowsher, D. (1984). Central pathways and mechanisms of pain sensation. In Holden, A. V. and Winlow, W. (eds), *Neurobiology of Pain*, Manchester University Press, Manchester, pp. 209–25

Boyadjieva, N. and Ovtcharov, R. (1987). Effect of some benzodiazepines on the secretion of thyrotropic hormone and prolactin under conditions of experimental stress, *Acta Physiol. Pharmacol. Bulg.*, **13**, 18–21

Boycott, B. B. (1965). Learning in the octopus, *Sci. Am.*, **212**, 42–50

Breward, J. and Gentle, M. J. (1985). Neuroma formation and abnormal afferent nerve discharges after partial beak amputation (beak trimming) in poultry, *Experientia*, **41**, 1132–4

Bunge, M. (1980). *The Mind–Body Problem. A Psychobiological Approach*, Pergamon, Oxford

Burgess, P. R. (1978). Peripheral nociceptive neurons: evidence for 'labelled' line versus 'common carrier' systems, *Neurosci. Res. Prog. Bull.*, **16**, 39–45

Burghardt, G. M. (1985). Animal awareness. Current perceptions and historical perspective, *Am. Psychol.*, **40**, 905–19

Carstens, E. E. (1987). Endogenous pain suppression mechanisms, *J. Am. Vet. Med.*, **191**, 1203–6

Casey, K. L. and Morrow, T. J. (1983). Supra-spinal pain mechanisms in the cat. In Kitchell, R. L. and Erickson, H. W. (eds), *Animal Pain: Perception and Alleviation*, American Physiological Society, Bethesda, MD, pp. 63–82

Cassano, G. F. (1983). What is pathological anxiety and what is not. In Costa, E. (ed.), *The Benzodiazepines: From Molecular Biology to Clinical Practice*, Raven, New York, pp. 287–93

Cassel, E. J. (1982). The nature of suffering and the goals of medicine, *New Engl. J. Med.*, **306**, 639–45

Chance, W. T. (1980). Autoanalgesia: opiate and non-opiate mechanisms, *Neurosci. Biobehav. Rev.*, **4**, 55–67

Chudler, E. H. and Dong, W. K. (1983). The assessment of pain by cerebral-evoked potentials. *Pain*, **16**, 221–4

Corda, M. G., Biggio, G. and Gessa, G. L. (1980). Brain nucleotides in naive and handling-habituated rats: differences in levels and drug sensitivity, *Brain Res.*, **188**, 287–90

Craig, K. D. (1984). Emotional aspects of pain. In Wall, P. D. and Melzack, R. (eds), *Textbook of Pain*, Churchill Livingstone, Edinburgh, pp. 153–61

Crawley, J. N., Skolnick, P. and Paul, S. M. (1984). Absence of intrinsical antagonistic actions of benzodiazepine antagonists on an exploratory model of anxiety in the mouse, *Neuropharmacol.*, **23**, 531–7

Crosby, E. C. and Showers, M. J. C. (1969). Comparative anatomy of the preoptic and hypothalamic areas. In Haymaker, E. A. and Nauta, W. J. H. (eds), *The Hypothalmus*, Charles C. Thomas, Springfield, IL, pp. 61–135

Davidson, J. M. and Davidson, R. J. (eds) (1980). *The Psychobiology of Consciousness*, Plenum, New York

Davidson, R. J. (1980). Consciousness and information processing: a bio-cognitive perspective. In Davidson, J. M. and Davidson, R. J. (eds), *The Psychobiology of Consciousness*, Plenum, New York, pp. 11–46

Dawkins, M. S. (1980). *Animal Suffering: The Science of Animal Welfare*, Chapman & Hall, London

Dawkins, M. S. (1986). The scientific basis for assessing suffering in animals. In Singer, P. (ed.), *In Defense of Animals*, Harper & Row, New York, pp. 27–40

Dennis, S. G. and Melzack, R. (1983). Perspectives on phylogenetic pain expression. In Kitchell, R. L. and Erickson, H. W. (eds), *Animal Pain: Perception and Alleviation*, American Physiological Society, Bethesda, MD, pp. 151–60

de Waal, F. B. M. (1982). *Chimpanzee Politics*, Jonathan Cape, London

Doré, F. Y. and Dumas, C. (1987). Psychology of animal cognition: Piagetian studies, *Pyschol. Bull.*, **102**, 219–33

Ebbesson, S. O. E. (1969). Brainstem afferents from the spinal cord in a sample of reptilian and amphibian species, *Ann. NY Acad. Sci.*, **167**, 80–101

Edwards, P. (ed.) (1972). *Encyclopaedia of Philosophy*, Macmillan, London

Ehlich, K. (1985). The language of pain, *Theor. Med.*, **6**, 177–87

Elton, D., Stanley, G. and Burrows, G. (1983). *Psychological Control of Pain*, Grune & Stratton, Sydney

Fields, H. L. and Basbaum, A. I. (1978). Brainstem control of spinal pain transmission neurons, *Annu. Rev. Physiol.*, **40**, 217–48

Fields, H. L. and Basbaum, A. I. (1984). Endogenous pain control mechanisms. In Wall, P. D. and Melzack, R. (eds), *Textbook of Pain*, Churchill Livingstone, Edinburgh, pp. 142–52

Fiorito, G. (1986). Is there 'pain' in invertebrates?, *Behav. Processes*, **12**, 383–8

Foltz, E. L. and White, L. I. (1962). Pain 'relief' by frontal cingulotomy, *J. Neurosurg.*, **19**, 89–100

Fraser, A. F. (1984/85). The behaviour of suffering in animals, *Appl. Anim. Behav. Sci.*, **13**, 1–6

Froehlich, J. W. (1984). Continuity between primate communication and human speech?, *J. Anthropol. Res.*, **40**, 597–602

Gallup, G. G. (1985). Do minds exist in species other than our own?, *Neurosci. Biobehav. Rev.*, **9**, 631–41

Gee, K. W., Yamamura, S. H., Roeske, W. R. and Yamamura, H. I. (1984). Benzodiazepine receptor heterogeneity: possible molecular basis and functional significance, *Fed. Proc.*, **43**, 2767–72

Gibson, T. E. (ed.) (1985). *The Detection and Relief of Pain in Animals*, BVA Animal Welfare Foundation, London

Gillman, M. A. and Lichtigfeld, F. J. (1985). A pharmacological overview of opioid mechanisms mediating analgesia and hyperalgesia, *Neurol. Rev.*, **7**, 106–19

Griffin, D. R. (1981). *The Question of Animal Awareness*, William Kaufman, Los Altos, CA

Griffin, D. R. (1984). *Animal Thinking*, Harvard University Press, Cambridge, MA

Guilbaud, G., Peschanski, M. and Besson, J. M. (1984). Experimental data related to nociception and pain at the supraspinal level. In Wall, P. D. and Melzack, R. (eds), *Textbook of Pain*, Churchill Livingstone, Edinburgh, pp. 110–18

Herman, B. H. and Panksepp, J. (1981). Ascending endorphin inhibition of distress vocalisation, *Science*, **211**, 1060–2

Holzman, A. D. (ed.) (1986). *Pain Management: A Handbook of Psychological Treatment Approaches*, Pergamon, New York

Iggo, A. (1984). *Pain in Animals*, Universities Federation for Animal Welfare, Hertfordshire

Izard, C. E. (1980). The emergence of emotions and the development of consciousness in infancy. In Davidson, J. M. and Davidson, R. J. (eds), *The Psychobiology of Consciousness*, Plenum, New York, pp. 193–216

Jakoubek, B. (1984). Analgesia induced by painful stimulation and/or anticipation of pain; different mechanisms are operating, *Physiol. Bohemoslov.*, **33**, 171–8

Jaynes, J. (1982). A two-tiered theory of emotions: affect and feeling, *Behav. Brain Res.*, **3**, 434–5

Johnson-Laird, P. N. (1983). *Mental Models. Towards a Cognitive Science of Language, Influence and Consciousness*, Cambridge University Press, Cambridge

Jones, M. W., Hodge, C. J., Apkarian, A. V. and Stevens, R. T. (1985). A dorso–lateral spinothalamic tract in cat, *Brain Res.*, **335**, 188–93

Jørum, E. (1988). Analgesia or hyperalgesia following stress correlates with emotional behaviour in rats, *Pain*, **32**, 341–8

Keele, K. D. (1957). *Anatomies of Pain*, Thomas, Oxford

Kitchell, R. L. (1987). Problems in defining pain and peripheral mechanisms of pain, *J. Am. Vet. Med.*, **191**, 1195–9

Kitchell, R. L. and Erickson, H. H. (1983). *Animal Pain: Perception and Alleviation*, American Physiological Society, Bethesda, MD

Kitchell, R. L. and Johnson, R. D. (1985). Assessment of pain in animals. In Moberg, G. P. (ed.), *Animal Stress*, American Physiological Society, Bethesda, MD, pp. 113–40

Kitchell, R. L., Naitoh, Y., Breazile, J. E. and Lagerwerff, J. M. (1962). Methodological considerations for assessment of pain perception in animals. In Keele, C. A. and Smith, R. (eds), *Assessment of Pain in Man and Animals*, Universities Federation for Animal Welfare, Hertfordshire, pp. 244–61

Kosterlitz, H. W., Lord, J. A. H., Paterson, S. J. and Waterfield, A. A. (1980).

Effects of changes in the structure of enkephalins and of narcotic analgesic drugs on their interactions with mu-receptor and delta-receptor, *Br. J. Pharmacol.*, **68**, 333–42

Kruger, L. and Rodin, B. E. (1983). Peripheral mechanisms involved in pain. In Kitchell, R. L. and Erickson, H. W. (eds), *Animal Pain: Perception and Alleviation*, American Physiological Society, Bethesda, MD, pp. 1–26

Lewis, A. (1980). Problems presented by the ambiguous word anxiety as used in psychopathology. In Burrows, G. D. and Davies, B. (eds), *Handbook of Studies on Anxiety*, Elsevier, Amsterdam

Lindley, R., Fellows, R. and Macdonald, G. (1978). *What Philosophy Does*, Open Books, London, pp. 53–89

Loew, F. M. (1981). Alleviation of pain: the researcher's obligation, *Lab. Anim.*, **10**, 36–8

Loew, F. M. (1987). The challenge of balancing experimental variables: pain, distress, analgesia and anesthesia. *J. Am. Vet. Med.*, **191**, 1193–4

McAllister, J. P. and Wells, J. (1981). The structural organisation of the ventral posterolateral nucleus in the rat, *J. Comp. Neurol.*, **197**, 271–301

McFarland, D. (ed.) (1985). *Oxford Companion on Animal Behaviour*, Oxford University Press, Oxford

Maiewski, S. F., Larscheid, P., Cook, J. M. and Mueller, G. P. (1985). Evidence that a benzodiazepine receptor mechanism regulates the secretion of pituitary β-endorphin in rats, *Endocrinology*, **117**, 474–80

Masserman, J. H., Wechklin, S. and Terris, W. (1964). 'Altruistic' behaviour in rhesus monkeys, *Am. J. Psychiatr.*, **121**, 584–5

Melzack, R. (1961). The perception of pain; reprinted in *Psychobiology; Readings from Scientific American*, pp. 299–307

Melzack, R. (1968). In Sills, D. L. (ed.), *International Encylopedia of the Social Sciences*, Macmillan, London, pp. 357–64

Melzack, R. (1973). *The Puzzle of Pain*, Basic Books, New York

Melzack, R. and Dennis, S. G. (1980). Phylogenetic evolution of pain expression in animals. In Kosterlitz, H. W. and Terenius, L. Y. (eds), *Pain and Society*, Elsevier, Amsterdam, pp. 13–26

Melzack, R. and Wall, P. D. (1965). Pain mechanisms: a new theory, *Science*, **150**, 971–9

Messenger, J. B. (1977). Prey capture and learning in the cuttle fish, *Sepia*. In Nixon, M. and Messenger, J. B. (eds), *The Biology of Cephalopods*, Academic, London, pp. 347–76

Moberg, G. P. (ed.) (1985). *Animal Stress*, American Physiological Society, Bethesda, MD

Molony, V. (1986). Assessment of pain by direct measurement of cerebro-cortical activity. In Duncan, I. J. H. and Molony, V. (eds), *Assessing Pain in Farm Animals*, Office for Official Publications of the European Communities, Luxembourg, pp. 79–85

Morton, D. B. and Griffiths, P. H. M. (1985). Guidelines on the recognition of pain, distress and discomfort in experimental animals and an hypothesis for assessment, *Vet. Rec.*, **116**, 431–6

Nathan, P. W. (1977). Pain, *Br. Med. Bull.*, **33**, 149–55

Nielsen, M., Braestrup, C. and Squires, R. F. (1978). Evidence for a late evolutionary appearance of brain-specific benzodiazepine receptors: an investigation of 18 vertebrate and 5 invertebrate species, *Brain Res.*, **141**, 342–6

Ninan, P. T., Insel, T. M., Cohen, R. M., Cook, J. M., Skolnick, P. and Paul, S. M. (1982). Benzodiazepine receptor-mediated experimental 'anxiety' in primates, *Science*, **218**, 1332–4

Nolan, A., Livingston, A., Morris, R. and Waterman, A. (1987). Techniques for comparison of thermal and mechanical nociceptive stimuli in the sheep, *J. Pharmacol. Meth.*, **17**, 39–49

OTA (Office of Technology Assessment) (1986). *Alternatives to Animal Use in Research, Testing and Education*, US Govt Printing Office, Washington, DC, p. 17

Panksepp, J. (1982). Toward a general psychobiological theory of emotion, *Behav. Brain Sci.*, **5**, 407–67

Parent, A. (1979). Anatomical organization of monoamine- and acetylcholin-esterase-containing neuronal systems in the vertebrate hypothalamus. In Morgane, P. J. and Panksepp, J. (eds), *Handbook of the Hypothalamus*, vol. 1, Marcel Dekker, New York, pp. 511–54

Pert, C. B., Aposhian, D. and Snyder, S. H. (1974). Phylogenetic distribution of opiate receptor binding, *Brain Res.*, **75**, 356–61

Piaget, J. (1967). *Biologie et Connaissance*, Gallimard, Paris

Pierau, F. K., Harti, G. and Taylor, D. C. M. (1985). Local administration of capsaicin does not deplete substance P in primary sensory neurones of the pigeon, *Pflugers Arch. Eur. J. Physiol.*, **403**, suppl., R60

Plutchik, R. (1980). A general psychoevolutionary theory of emotion. In Plutchik, R. and Kellerman, H. (eds), *Emotion, Theory, Research and Experience*, Academic, Orlando, FL, pp. 3–33

Premack, D. (1984). Comparing mental representation in human and non-human animals, *Social Res.*, **51**, 983–99

Price, D. D., Hayes, R. L., Ruda, M. and Dubner, R. (1978). Spatial and temporal transformations of input to spinothalamic tract neurons and their relation to somatic sensations, *J. Neurophysiol.*, **41**, 933–47

Rachlin, H. (1985). Pain and behaviour, *Behav. Brain Sci.*, **8**, 43–83

Ralston, H. J. (1982). The neuronal and synaptic organisation of the ventrobasal thalamus in rat, cat and monkey, *First World Congress of IBRO*, Milan

Ralston, H. J. (1984). Synaptic organisation of spinothalamic tract projections to the thalamus, with special reference to pain, *Adv. Pain Res. Ther.*, **6**, 183–95

Reich, W. T. (ed.) (1978). *Encyclopedia of Bioethics*, Free Press, New York

Richards, J. G. and Mohler, H. (1984). Benzodiazepine receptors, *Neuropharmacol.*, **23**, 233–42

Riley, A. L., Zellner, D. A. and Duncan, H. J. (1980). The role of endorphins in animal learning and behaviour, *Neurosci. Biobehav. Rev.*, **4**, 69–76

Ristau, C. A. (1983). Symbols and indication in apes and other species. Comment on Savage-Rumbaugh *et al.*, *J. Exp. Psychol.*, **112**, 498–507

Roitblat, H. L., Bever, T. G. and Terrace, H. S. (eds) (1984). *Animal Cognition*, Lawrence Erlbaum, London

Romanes, G. J. (1888). *Mental Evolution in Man.* Kegan-Paul, Trench, Trubner, London

Rust, M., Keller, M., Gessler, M. and Zieglgansberger, W. (1984). Endorphinergic mechanisms in pregnancy-specific pain adaptation, *Anaesthetist,* **33**, 452

Sanford, J., Ewbank, R., Molony, V. , Travenor, W. D. and Uvarov, O. (1986). Guidelines for the recognition and assessment of pain in animals, *Vet. Rec.,* **118**, 334–8

Savage-Rumbaugh, E. S., Pate, J. L., Lawson, J., Smith, S. T. and Rosenbau, S. (1983). Can a chimpanzee make a statement?, *J. Exp. Psychol.,* **112**, 457–92

Schumacher, R. and Velden, M. (1984). Anxiety, pain experience and pain report: a signal detection study, *Percept. Motor Skills,* **58**, 339–49

Schaffer, J. A. (1978). Pain and suffering. Philosophical perspectives. In Reich, W. T. (ed.), *Encyclopedia of Bioethics,* Free Press, New York, pp. 1181–4

Sills, D. L. (ed.) (1968). *International Encylopedia of the Social Sciences,* Macmillan and Free Press, London

Spector, W. G. (1980). *An Introduction to General Pathology,* Churchill Livingstone, Edinburgh

Stanley, T. H. (1987). New developments in opioid drug research for alleviation of animal pain, *J. Am. Vet. Med.,* **191**, 1252–3

Stephens, D. B. (1988). A review of experimental approaches to the analysis of emotional behaviour and their relation to stress in farm animals, *Cornell Vet.,* **78**, 155–77

Sternbach, R. A. (1974). *Pain Patients,* Academic, New York

Tannenbaum, J. and Rowan, A. N. (1985). Re-thinking the morality of animal research, *Hastings Center Rep.,* **5**, 32–43

Terman, G. W., Shavit, Y., Lewis, J. W., Cannon, J. T. and Liebeskind, J. C. (1984). Intrinsic mechanisms of pain inhibition: activation by stress, *Science,* **226**, 1270–7

Terrace, H. S. (1984). Animal cognition. In Roitblat, H. L., Bever, T. G. and Terrace, H. S. (eds), *Animal Cognition,* Lawrence Erlbaum, London, pp. 7–28

Teitz, W. J. and Hall, P. (1977). In Swenson, M. J. (ed.), *Dukes' Physiology of Domestic Animals,* 9th edn, Cornell University Press, Ithaca, NY, pp. 671–85

Truex, R. C., Taylor, M. J., Smythe, M. Q. and Gildenberg, P. L. (1970). Lateral cervical nucleus of cat, dog and man, *J. Comp. Neurol.,* **139**, 93–104

Vaswani, K., Tejwani, G. A. and Mousa, S. (1983). Stress-induced differential intake of various diets and water by rat: the role of the opiate system, *Life Sci.,* **32**, 1983–96

Vierck, C. J. and Cooper, B. Y. (1984). Guidelines for assessing pain reactions and pain modulation in laboratory animal subjects, *Adv. Pain Res. Ther.,* **6**, 305–22

Vogel, J., Beer, B. and Clody, P. (1971). A simple reliable conflict procedure for testing antianxiety agents, *Psychopharmacol.,* **21**, 1–7

Wall, P. D. (1978). The gate control theory of pain mechanisms: a re-examination and re-statement, *Brain,* **101**, 1–18

Wall, P. D. (1979). On the relation of injury to pain, *Pain*, **6**, 253–64
Wall, P. D. and Melzack, R. (eds) (1984). *Textbook of Pain*, Churchill Livingstone, Edinburgh
Warnock, M. (1985). Law and the pursuit of knowledge, *Conquest*, **175**, 1–7
Watkins, L. R., Suberg, S. N., Thurston, C. L. and Culhane, E. S. (1986). Role of spinal cord neuropeptides in pain sensitivity and analgesia: thyrotropin-releasing hormone and vasopressin, *Brain Res.*, **362**, 308–17
Weinstein, J., Hough, L. B. and Gogas, K. R. (1988). Cross-tolerance and cross-sensitization between morphine analgesia and naloxone-sensitive and cimetidine- sensitive stress induced analgesia, *J. Pharmacol. Exp. Ther.*, **244**, 253–8
Whiting, H. P. (1955). Functional development in the nervous system. In Waelsch, H. (ed.), *Biochemistry of the Developing Nervous System*, Academic, New York, pp. 85–102
Williams, R. B. and Eichelman, B. (1971). Social setting: influence on the physiological response to electric shock in the rat, *Science*, **174**, 613–14
Willis, W. D. (1984). The origin and destination of pathways involved in pain transmission. In Wall, P. D. and Melzack, R. (eds), *Textbook of Pain*, Churchill Livingstone, Edinburgh, pp. 88–99
Willis, W. D. and Chung, J. M. (1987). Central mechanisms of pain, *J. Am. Vet. Med.*, **191**, 1200–2
Wilson M. (1985). What is this thing called 'pain'? — the philosophy of science behind the contemporary debate, *Pacific Phil. Qt.*, **66**, 227–67
Wolfle, T. L. (1987). Control of stress using non-drug approaches, *J. Am. Vet. Med.*, **191**, 1219–21
Wright, E. M., Marcella, K. L. and Woodson, J. F. (1985). Animal pain: evaluation and control, *Lab. Anim.*, **9**, 20–36
Wüster, M., Rubini, P. and Schultz, R. (1981). The preference of putative pro-enkephalins for different types of opiate receptors, *Life Sci.*, **29**, 1219–27
Young, J. Z. (1961). Learning and discrimination in the octopus, *Biol. Rev.*, **36**, 32–96
Zayan, R. (1986). Assessment of pain in animals: epistemological comments, In Duncan, I. J. H. and Molony, V. (eds), *Assessing Pain in Farm Animals*, Office for Official Publications of the European Communities, Luxembourg, pp. 1–15
Zimmerman, M. (1983). Ethical guidelines for investigations of experimental pain in conscious animals, *Pain*, **16**, 109–10
Zimmerman, M. (1984). Ethical considerations in relation to pain in animal experimentation. In Bankowski, Z. and Howard-Jones, N. (eds), *Biomedical Research Involving Animals*, CIOMS, Geneva, pp. 132–9
Zimmerman, M. (1986). Behavioural investigations of pain in animals. In Duncan, I. J. H. and Molony, V. (eds), *Assessing Pain in Farm Animals*, Office for Official Publications of the European Communities, Luxembourg, pp. 16–27

4. Methods and Practices of Animal Experimentation

Erik Millstone

'. . . the best model for human cancer research is Herringus rufus *. . . The histology of the tumours is identical to that in man but, if necessary it may be different. Although the red herring lives only two years, correction factors can be applied such that a herring aged 18 months is equivalent to a human being aged 20–70 years: the incidence of tumours is then found to be the same as, or different from, that in man.'*

Anon (1974). Animal models in cancer research, *Lancet*, **ii**, 1506

INTRODUCTION

Some people argue that it can never be legitimate to experiment on animals, but that is not my view. While I reject the view that the lives of animals have no value, I do unapologetically ascribe a greater value to the lives of humans than to those of animals. While many animal experiments may be cruel, unnecessary and undesirable, well-conducted relevant animal experiments which contribute to a significant reduction in human suffering (or even a reduction in the probability of human suffering) provide the lesser of a set of evils, and may therefore be ethically legitimate. The central issue, however, which this essay will address concerns the extent to which information about animals obtained from experiments can usefully be extended to apply to humans. If animals are poor models for humans, then many experiments are wasteful, unnecessary and uninformative, and the cruelty is entirely gratuitous and indefensible.

It is not possible in this context to review the entire range of medically related animal experimentation, but several examples from the field of toxicology may be sufficient to provoke serious doubts about the validity, relevance and value of animal experiments as currently practised. Toxicology is that activity to which we turn when we want to know if environmental chemicals are safe or toxic. Animal tests currently play a central role in toxicology as one tries to predict the safety of, or toxicity of, these chemicals for humans. It may well be ethically legitimate to use animal tests, but only if potentially

72

significant hazards to humans from the use of chemicals can be reliably detected by the use of animal tests.

CRITERIA FOR ANIMAL STUDIES

At least four conditions have to be satisfied if animal tests are to be useful, not just in the shallow sense of enabling chemical and pharmaceutical companies to negotiate their way over governmental regulatory hurdles, but in the substantial sense of informing us about the likely effects of compounds on human health. First, there has to be sufficient similarity between the anatomy, physiology and metabolism of experimental animals and those of humans to provide a basis for reliable extrapolation. Secondly, there has to be a satisfactory match between the lifestyles of the laboratory animals and the diverse circumstances of the human population whose health the testers wish to model and protect. Thirdly, there need to be well-defined systematic and valid rules for extrapolating from the results of animal tests to conclusions about humans; and fourthly, those rules of extrapolation need to be reliably and diligently followed by official regulatory institutions. Unfortunately, in relation to important parts of toxicology, these four conditions are frequently not satisfied. It does not follow from this, however, that their satisfaction is unattainable.

Before examining these four conditions it is important to appreciate that arguments concerning the extrapolative validity of animal tests need to be understood in part by reference to their social contexts. For example, in the USA, where regulatory regimes are relatively strict, the results of animal tests are being used as grounds for restricting or banning chemicals. Consequently, those Americans who argue from an industrial perspective, such as Efron (1984) and the American Council on Science and Health (ACSH 1984), are extremely critical of the validity of animal tests. On the other hand, Epstein (1978), for example, arguing from the point of view of public protection, defends the relevance and validity of extrapolations from animal data. In the UK, by contrast, where regulatory regimes are often less restrictive, animal test data are used to provide clearance for commercial compounds, and industrial representatives defend the value of animal tests against criticisms from public-interest groups (Gillespie *et al.* 1979, Millstone 1986, Conning 1985).

Context is important in at least one further respect. When we are concerned with diminishing risks to health, and when we are trying to decide which industrial products and processes to permit and which to forbid, we can divide all the cases into two broad classes. The first class includes all of those cases where a health benefit is expected from the consumption of the material. Examples include pharmaceutical products having a therapeutic effect, and food preservatives which inhibit bacterial spoilage and consequent food poisoning. For all examples in this category, we might incur health risks by failing to consume the compounds, even though there may also be possible risks arising from their consumption. Some of our most powerful antibiotics like chloramphenicol may increase our vulnerability to aplastic anaemia, but most of us would choose to take that risk rather than die tonight of typhoid fever. In cases falling within this group we want to make a decision by comparing two different risks with each other. In the other class of cases, including for example artificial food colours and cosmetic chemicals, their safety is important, but if we avoid them completely we will be taking no risks with our health.

SPECIES DIFFERENCES

As to the first condition, there are evidently both similarities and differences between humans and laboratory animals. What matters is the significance of those features which we have in common with, and those which differentiate us from, other animals. While all mammals have numerous common features, humans are not rodents. Rodents, for example, have forestomachs while humans do not. This is important because, for example, the widely used antioxidant food additive butylated hydroxyanisole (BHA, E320) causes cancers in the forestomachs of both rats and hamsters when their diets include 2% of BHA (Ito *et al.* 1985). Since humans do not have forestomachs, and consume BHA at far lower doses, the significance of the rodent data is hard to assess (Millstone and Abraham 1988, pp. 100–3). Human beings and laboratory rodents differ in numerous ways and we are not yet in a position to be able to explain the significance of most of these differences. That fact by itself might incline us to caution when it comes to extrapolations; however, there is evidence which suggests not just that our inferences from animals to humans are unreliable, but that they may be actively misleading. For example, we have direct evidence from occupational epidemiology that β-naphthylamine causes cancer in humans, but most strains of rats are not similarly vulnerable (Efron 1984, p. 190). In a case like this, reliance on a rat carcinogenicity study to approve commercial use would entail unintended and unacceptable risks.

The thalidomide tragedy played an important role in tightening the testing requirements for new drugs, but it also provides an example of substantial inter-species differences. A dose of one milligram per kilogram of body weight per day ($1 \text{ mg kg}^{-1} \text{ d}^{-1}$) in humans is sufficient to produce a foetotoxic effect. In the rabbit the effective dose must be 30 mg kg^{-1} d^{-1} or more, while the Wistar rat is unaffected by doses of up to 4000 mg kg^{-1} d^{-1} (Kalter 1968). Similarly, recent research has indicated that while both rats and humans metabolise the artificial sweetener aspartame into phenylalanine, humans do so at a rate which has an influence on a crucial aspect of brain biochemistry 60 times greater than that characteristic of rats (Wurtman and Maher 1987). On the other hand, miners have long been able to benefit from the fact that canaries are far more vulnerable to carbon monoxide than are humans. So while we should be wary of placing too great a reliance on animal studies, they should not be dismissed, but rather improved.

When discussing the (in)adequacy of animal tests for the safety of pharmaceutical products, Litchfield (1962) explains that the differences between species are '. . . so pronounced that a system utilizing drugs could be devised to prove that an experimental animal was a rat and not a dog or a man. There are . . . easier ways of making this distinction, but the point is that drugs can produce effects which are unique to a given species'. There are numerous such examples where we know that chemicals cause adverse effects in humans even though no counterpart evidence can be found in laboratory animals. We have so far failed to find any animal model for exfoliative dermatitis which, for example, some drugs can cause in human subjects (Litchfield 1962). There is evidence from double-blind placebo challenge studies to show that some foods and food additives provoke severe symptons of acute intolerance in some human subjects, but these symptons too are not reproducible in laboratory animals (Egger *et al.* 1983, 1985, Supramaniam and Warner 1986).

Until recently, no systematic attempt had been made to estimate precisely the extrapolative relevance of animal studies to human health. In 1983, however, Salsburg published the first quantitative estimate of the human relevance of long-term mouse and rat feeding studies for chemical carcinogenicity (Salsburg 1983). The virtue of Salsburg's approach was that he directly addressed the issue of the criteria against which the relevance of animal studies needs to be judged. The value of animal studies should be established by comparing animal data with the results of studies in human epidemiology and with clinical data. Laboratory tests have enabled us to identify over 380 different compounds which can provoke tumours in rodents, but we have no idea which of them cause cancer in humans. Long-term toxicity in humans, such as carcinogenicity, is very hard to study epidemiologically, and we have so far only been able to identify about 26 compounds or groups of chemicals which definitely cause human cancers. With the exception of tobacco smoke (which for these purposes counts as one group of chemicals), human exposure to these carcinogens is now tightly controlled and consequently we are profoundly ignorant of the causes of almost all the non-tobacco-related cancers. Toxicologists are therefore trying to deal with problems which are both difficult and urgent. This partly accounts for why it is that animal tests continue to be required and conducted despite their limited value.

Salsburg (1983) first reviewed the data on 19 of the 26 groups of compounds and asked the question: 'Have these compounds been shown to be carcinogenic in long-term rodent feeding studies?' To examine the relevance of feeding studies he initially set aside those seven compounds which are only known to cause cancer through inhalation rather than by being eaten. For those 19 compounds he concluded that the animal studies got it right approximately 37% of the time. Including the inhalation carcinogens raised the correlation to about 46%. Salsburg concludes therefore that ' . . . the lifetime feeding study of mice and rats appears to have less than a 50% probability of finding known human carcinogens. On the basis of probability theory, we would have been better off to toss a coin' (Salsburg 1983, p. 64).

Salsburg's findings have been disputed. His work has provoked some further testing, often at doses higher than those previously used. The most optimistic protagonist has estimated that animal feeding studies have so far correctly identified known human carcinogens 75% of the time (Haseman and Salsburg 1983, Salsburg and Haseman 1984). A correlation of 75% for known human carcinogens is better science than 37%, but it remains a very poor basis for policy-making, especially if we bear in mind that the correlation between human studies and rodent tests should be at its highest for known human carcinogens as these are the problems which have received most attention and have been most thoroughly tested. We should expect the correlation to be weaker for all the other so-far-undetected human carcinogens and for all other hazards.

Salsburg's findings are, furthermore, open to a variety of interpretations. One response might be to conclude that rodents are too different from humans ever to provide an adequate model. Salsburg's view appears to be that the problems lie not with the inter-species differences, but with the standard methods and procedures that are currently being used. Salsburg is hopeful that by conducting rodent tests differently we may refine the correlations to the point where the models become relevant and useful, but I see few grounds for his optimism (Salsburg 1985). Current test protocols are criticised on numerous grounds but most commonly because the doses used are often very high, and this brings us to the second condition.

THE LABORATORY ENVIRONMENT AND RULES FOR EXTRAPOLATION

The human population is large, complex and diverse. Levels of exposure to putatively hazardous chemicals may vary by several orders of magnitude. The standardised international protocol for long-term feeding studies involves adding the test compound to the diets of rodents at three dose levels, and comparing them to control animals receiving an uncontaminated diet. The highest dose is normally what has been termed the 'maximum tolerated dose' (MTD), and the two lower doses are fractions of the MTD. At each dose level 50 animals of each sex are used. The MTD is defined as the highest dose at which the body weight of the test animals does not fall more than 10% below that of the concurrent controls. In practice the MTDs used in tests are often substantially greater, as a proportion of body weight, than plausible levels of human exposure. Consequently many commentators criticise MTD testing, and call for 'more realistic' lower dose levels.

The rationale for MTD testing, however, is an important one. Because as few as 400 short-lived animals are being used to model a human population of many millions with life expectancies as high as 70 years or more, high doses are supposed to compensate for population size and life-expectancy differences. The truth is that we just do not know whether or not these factors will compensate for each other. If, however, we move away from MTD testing towards a reliance on far lower doses, without a massive and very costly corresponding increase in the number of animals used, then the tests will become even less sensitive, while our underlying concern is that they are already too insensitive. Given the numbers of animals used in long-term feeding studies and the statistical methods used to analyse their results, we can be at best 95% confident of detecting a carcinogen which caused cancer in approximately 15% of the animals. If a carcinogen is any weaker than that, it becomes increasingly likely that animal tests conducted in accordance with current protocols will fail to detect its toxicity.

All of these concerns are compounded by the fact that, while the human population is diverse in terms of genetics, health, diets, exposures and environments, laboratory animal populations are often genetically uniform and free of environmental pathogens. They have access to unlimited quantities of food despite being confined in small cages which preclude much exercise. When the safety or efficacy of compounds are tested, this is almost invariably accomplished by adding one test substance at a time to animal diets, while patients frequently receive several drugs at any one time. Food additives, moreover, are consumed in many complex combinations. The possible effects of these chemical cocktails are not replicated in animal studies.

There are consequently good grounds for doubting that the circumstances of laboratory animals can satisfactorily model the real circumstances of many humans who may be at risk.

The problems which we face when trying to extrapolate from the results of laboratory experiments on animals to plausible patterns of human consumption are very complex. The chain of reasoning starts from a small but homogeneous group of animals receiving relatively high doses and moves initially to larger but still homogeneous populations of the same species of animals receiving lower doses. It is at the next stage where we seek to extrapolate between species when a conclusion is drawn about a large but homogeneous and healthy population of people. Further steps are required, however, if we are to draw conclusions about large human populations which are heterogeneous in respect of age, health, dosage and exposures to other chemicals.

Nobody has yet attempted to quantify the magnitude of the compound uncertainties inherent in each of these three stages, but we have one published estimate of the uncertainties inherent in the first two. A team of American scientists, starting from data on the extent to which the artificial sweetener saccharin caused cancer in the bladders of male rats, eventually concluded that ' . . . over the next 70 years, the expected number of cases of human bladder cancer in the USA resulting from daily exposure to 120 mg saccharin might range from 0.22 to 1 144 000' (Wilkinson 1983). If this estimate is correct, and I know of no challenge to it, then we must conclude that the uncertainty in quantitative extrapolations may be as large as, or even larger than, seven orders of magnitude. When supposedly 'acceptable daily intakes' (ADIs) of food additives are set by the authoritative Joint Expert Committee of the UN Food and Agriculture Organisation and the World Health Organisation, they normally rely on a safety factor of 100. The ADI is normally set at one-hundredth of a dose level, per unit weight of the animal, at which no adverse effects are observed. If the previous estimate is correct then at best the 'safety factor' is less than the cube root of the uncertainty.

In the face of all the uncertainties, it is perhaps understandable that an anonymous author was prepared to joke in the pages of the *Lancet* and suggest that ' . . . the best model for human cancer research is *Herringus rufus* . . . The histology of the tumours is identical to that in man but, if necessary it may be different. Although the red herring lives only two years, correction factors can be applied such that a herring aged 18 months is equivalent to a human being aged 20–70 years: the incidence of tumours is then found to be the same as, or different from, that in man' (Anon 1974).

The failure to satisfy the first of our original two conditions entails that the third condition cannot yet be satisfied. We do not have sufficient information with which to construct rules of inference from animals to humans. An attempt has been made *post facto* to construct a set of rules of inference to account for the regulatory advice of expert committees, but from our point of view that amounts to an attempt to put the cart before the horse (Chu *et al.* 1981).

The failure to satisfy the first three conditions entails, moreover, the failure of the fourth. In the absence of any consensual rules we cannot even enquire about their putative validity, nor whether they might be applied consistently. It is important, however, to examine the ways in which expert committees and regulatory institutions handle toxi-cological data when the implications of those data are controversial or ambiguous. It is difficult to provide reliable generalisations about the conduct of many committees in numerous different regimes dealing with a very large number of compounds and patterns of exposure. There are, however, four main types of circumstance which need to be distinguished.

The two extremes are relatively straightforward while the two intermediate cases pose the difficulties. At the safe end of the spectrum we may find a compound which has been tested on animals without any adverse effects having being found, and where there is no epidemio-logical or clinical evidence of a human hazard. Regulatory agencies invariably treat such compounds as acceptably safe. The opposite extreme would be a chemical which is demonstrably toxic to both animals and humans, and which is therefore subject to restrictions or eliminated altogether. Cases of those two sorts are relatively straight-forward and need not detain us; the two remaining possibilities are, however, especially problematic.

REGULATORY POLICY DECISIONS

There is sometimes evidence that a compound is hazardous to laboratory animals even though there is no corroborative evidence that it is harmful to humans. In some cases the dose at which animals suffer is relatively far larger than any plausible level of human exposure, but we cannot just assume that rodents and humans would be vulnerable in similar ways at similar doses. When we are considering a putative chronic toxic hazard to humans, it is often true that there is no epidemiological evidence to reinforce the indications from animal studies. Frequently that fact needs to be taken with a medium-sized pinch of salt. It is often important to ask whether anyone has conducted any epidemiological studies to search for such evidence, and whether the methods and resources of modern epidemiology are sufficiently powerful to detect any adverse effects that may be occurring. Often negative answers have to be given to both those questions. In relation to a narrow range of examples, generally drawn from pharmacology, epidemiologists and/or clinicians are able to accumulate at least some confirmatory circumstantial evidence, but it is rarely decisive.

When dealing with examples falling into this broad group the question we need to ask is: 'Who is receiving the benefit of the doubt?' If the public (comprising consumers, workers and third parties) is to be protected then we would expect regulatory authorities to treat the chemical as hazardous. If, on the other hand, the benefit of the doubt were being awarded to industry then the regulators would treat the chemical as safe until direct human evidence of a hazard can be provided. We can find examples fitting this latter pattern amongst food additives, agricultural pesticides and pharmaceutical products. For example, the evidence, referred to above, that the antioxidant BHA causes cancer in the forestomachs of rodents was sufficient to persuade the Japanese authorities to ban it, while in the UK, the EEC and the USA its use continues to be approved.

During the mid-1970s evidence from animal experiments emerged indicating that the synthetic red colouring amaranth (Red No. 2, E123) might cause spontaneous miscarriages, birth defects and cancer (Verrett and Carper 1974, Doyal *et al.* 1983, Millstone 1986, pp. 121–4). The response in the USA, the USSR, Greece, Yugoslavia, Norway, Finland and Austria was to ban it, but it remains in use in the UK and approximately 63 other countries (Millstone and Abraham 1988, pp. 34–7). Similarly, on the basis of identical toxicological dossiers, the US Food and Drug Administration judged the two related pesticides aldrin and dieldrin to be carcinogenic, while the British authorities passed them as acceptably safe (Gillespie *et al.* 1979). Over the last 20 years, evidence has progressively accumulated indicating that the anticonvulsive drug sodium valproate is teratogenic in mice, rats and rabbits. The first fragment of clinical data on adverse teratogenic effects on humans emerged in August 1980, and several further fragments have surfaced (Dalens *et al.* 1980, Gomez 1981). It would be illusory to pretend that we can be certain whether or not this drug is entirely safe for epileptics who are reproductively active. We can therefore readily appreciate why a group of Scandinavian doctors, writing to *The Lancet* in 1982, called for ' . . . a critical revaluation of teratogenicity tests and revisions of their design' (Sune Larsson *et al.* 1982, Baker and Davey 1970).

A further illustration of the kinds of problems encountered can be found in the debate about the safety or toxicity of lead in petrol. As Collingridge and Reeve (1986) have pointed out, the Report of a Department of Health and Social Security (DHSS) Working Party on

Lead in the Environment was '. . . written by a panel of scientists, none of whom were experienced in animal studies or biochemistry . . . and . . . rejected all animal studies as irrelevant to an understanding of the medical effects of lead on humans'. The Committee disregarded evidence from biochemical and animal studies which implied that lead at levels at or below 5 micrograms per 100 millilitres (5 µg/100 ml) of blood could adversely affect children's health (Collingridge and Reeve 1986, p. 47). They selected, however, from amongst the human epidemiological data those results which appeared to indicate the absence of any hazard (DHSS 1980). By contrast, the Conservation Society (1980) drew attention to the animal data while discounting the failure of some epidemiologists to detect significant adverse effects. The disparity in the two approaches has been quantified by Reeve, who reports that 'Of the 276 papers cited in the Conservation Society report and the 125 by Lawther, only thirty were cited by both parties' (Collingridge and Reeve 1986). This illustrates the point that different interest groups will often place divergent and conflicting emphases on animal and epidemiological evidence so as to accord with their institutional commitments.

While Collingridge and Reeve conclude that the disparity between the conflicting reports shows that there are no good grounds either for or against regulatory action, I would argue that if a null result in an animal test is taken as evidence of safety then evidence of adverse effects should always be interpreted as *prima facie* evidence of hazard. As Salsburg acknowledges, we ought to judge the adequacy of animal tests by comparing their results with human data. To continue to permit the use of a compound which is toxic to animals in the absence of adequate evidence of a human hazard might seem reasonable until we appreciate that inconsistent standards are entailed. We cannot dismiss the relevance of adverse animal findings when they appear to indicate a hazard while choosing to rely on the absence of evidence of a hazard to animals as grounds for judging a chemical safe. We cannot have, or at any rate we should try not to have, it both ways. If we rely on animal studies to provide evidence of safety we should accept adverse findings however unwelcome they may be. If the public is to be protected, and if animals are not to be pointlessly sacrificed to industrial and administrative convenience, then the results of animal studies must be interpreted consistently, and the benefit of the doubt should be given to the public.

The remaining class of intermediate cases would be those in which there is direct clinical or epidemiological evidence of a hazard to humans and where the animal tests have been conducted but there is no confirmatory evidence that the animals are reacting adversely. Such lack of agreement between human and animal data admits of several interpretations. The response which is scientifically most defensible, and the one which is to be preferred on social grounds, would be to conclude that we have identified significant respects in which animal tests are deficient, and a prudent policy of public protection would entail restricting or banning the toxic agent and deciding not to rely solely on comparable animal data in similar contexts on future occasions. The alternative response would be to insist on the apparent implications of the animal studies and conclude that the compound was safe and dismiss the human data as variously coincidental, artefactual or even hysterical, but at any rate mistaken.

The debate about a leukaemia cluster in the neighbourhood of a nuclear installation provides a relevant example. In November 1983 a Yorkshire Television documentary drew attention to a cluster of cases of childhood leukaemia in the vicinity of Sellafield. The official response was to establish a Committee of Inquiry under the chairmanship of Sir Douglas Black, a former president of the Royal College of Physicians, with responsibility to advise on whether or not the local plant operated by British Nuclear Fuels Ltd (BNFL) could have been responsible. Despite their eminence, the members of that Committee did not have all the relevant expertise, and therefore they subcontracted some of their inquiry to the National Radiological Protection Board (NRPB). The NRPB report provided conclusions which were based *inter alia* on extrapolations from the results of animal studies. The NRPB considered information about the metabolism of, and the form of the dose–response curves for, radionuclides in laboratory animals, and combined them with estimates of exposure levels to the local Cumbrian community, and concluded that the likely rate of childhood leukaemia in that community should be approximately 1/400th of that which was observed epidemiologically, while the epidemiological evidence indicated that the incidence of leukaemia in Seascale was 10 times the national average (NRPB 1984).

The contrast between the high rate of incidence which was reported epidemiologically and the low rate predicted by the NRPB was interpreted by both the NRPB and the Black Committee as indicating that the BNFL plant could not have been the cause. Their logic is distinctly idiosyncratic. It implies that if a very low level of leukaemia had been observed which equated with their prediction then they might have accepted that the plant may have been causally responsible, but with an observed cancer rate 400 times higher they acquit the plant of all responsibility. On this logic, the higher the observed rate, the less likely were emissions from the plant to have been responsible. This eccentricity can be explained and avoided as soon as we appreciate that the NRPB's estimate of the likely rate of leukaemia was based on an extrapolation from animal data. This extrapolation was very complex, consisted of many stages, and depended on numerous assumptions. As Russell Jones (1984) has pointed out, many of the figures which entered into the NRPB's calculation were presented as if they were measurements when in fact they were often ' . . . no more than . . . theoretical prediction[s] based on a large number of assumptions, any of which may be subsequently discredited'.

Crouch has reviewed many of the assumptions and corresponding uncertainties and has shown that the extrapolations from the animal models may plausibly and readily be modified to generate a prediction corresponding to observed epidemiological data (Crouch 1986, 1987). The crucial issue in this context, however, is the fact that it provides an example of official British agencies accepting the implications of animal studies in preference to direct epidemiological data. Whatever else it might have done, the NRPB report certainly gave the benefit of the doubts to BNFL by preferring animal data (albeit heavily interpreted) which minimised the apparent risks while downplaying the significance of direct epidemiological evidence of a hazard to human children.

A further example of preferring animal data to epidemiology occurs in the field of food additives and the putative acute toxicity of some compounds. There is direct epidemiological and clinical data which indicate that a small but so far undetermined portion of the population reacts adversely to some colours, preservatives, antioxidants and flavour enhancers (Royal College of Physicians 1984, Egger *et al.* 1985, Millstone 1986, pp. 106–113). The symptoms include, but are not exhausted by, hyperactivity, asthma and eczema. Conning (1983), who was then the Director of the British Industrial Biological Research Association which conducts toxicological studies under contract for the food industry, cites the fact that animals do not exhibit corresponding symptoms as grounds for doubting the reality of the interaction. There are, moreover, many other cases where known human toxicity has yet to be replicated in an animal model (Kowlczyk 1987). This lack of correspondence between human experience and animal models merely undermines the credibility of the animal tests, and in no way diminishes the human suffering.

CONCLUSIONS

The main conclusions to be drawn from this discussion are that the use of animals in toxicological studies does not provide a reliable basis for extrapolation to human health. While animal tests enable us to learn how commercial chemicals affect mice, rats, rabbits and guinea-pigs, we know very little about their probable effects on humans. I have argued moreover that, in Britain at any rate, the criteria by which the results of animal studies are judged are often inconsistent and commercially biased. If a compound is tested on animals and no adverse effects are found then this is being used to provide grounds for the conclusion that the compound is acceptably safe. If tests indicate that the compound is toxic to animals this is often interpreted as showing only that it would be unacceptable for use on animals, but not that its use by humans should be tightly restricted. Even where human epidemiological and clinical data contradict the apparent implications of some animal studies, some scientists and institutions are ascribing greater credence to animal studies than to human data. In other words, the animal studies are believed when they provide the answer that is required and ignored when they indicate an unwanted conclusion. In these circumstances it is impossible to defend a continuation of current practices.

There are two key problems. The first is the extensive uncertainty which afflicts inter-species extrapolations and the second is the institutionalised exploitation of those uncertainties. If the use of animals in toxicological experiments is to be useful and legitimate then both problems need to be addressed. They are, however, connected because, as Conning has explained, toxicology developed as a response to the demands of regulators and not as a science directed to understanding the biochemical impacts of environmental chemicals (Conning 1985).

Animal tests are not always being used to protect potential victims, but as part of a ritual enabling industry to negotiate its way through regulatory hurdles. Mice and rats are not used because their selection makes biological sense. They are used despite their lack of biological relevance because they are ' . . . readily available, and easy to handle and house' (Baker and Davey 1970, Calabrese 1982). These rodents are small, cheap, standardised, with a brief life span, and there is no strong lobby seeking to protect their interests. This state of affairs is neither desirable nor inevitable. We could and should initiate a research programme which will enable us to reduce the extrapolative uncertainties. Following Salsburg's example, we should compare the results of animal tests with the knowledge which we can obtain by epidemiological and clinical studies, and then calculate the extrapolative validities of existing test procedures. We would then be in a position to identify the areas where the correlations are poorest and starting in those areas develop alternative and preferably non-animal tests which are demonstrably more relevant and valid. As well as comparing the results of animal tests with epidemiological and clinical studies we need to supplement these comparisons with metabolic and pharmacokinetic information so that we can compare not just outcomes but processes too. We might then find, for example, that while the rat kidney is a good model for the human kidney, rabbits have lungs which are closer to ours than are those of the rat. We might also find that human kidney and lung cells in culture are the best model of all. It is an indication of the inadequacies and complacency of the toxicological establishment that these questions have never yet been properly addressed.

As and when we have test procedures, and rules for interpreting the data, which are consistent and demonstrably more relevant than current procedures, then we will have greater confidence that public health can be adequately protected, and that such animal tests as are conducted are informative and justifiable. If and when that stage is reached there will be far less scope for vested interests to select and interpret data in a biased fashion. In the meantime, there are urgent decisions which have to be made. It would be unreasonable to suspend the introduction of all new compounds until the toxicological uncertainties have been substantially diminished. In the interim we should not abandon all animal tests, because that might entail taking unacceptable risks with public health, but rather we should tighten up the criteria by which animal data are interpreted so as to ensure that it is consumers, workers and the public who receive the benefit of the remaining doubts.

REFERENCES

ACSH (American Council on Science and Health) (1984). *Of Mice and Men: The Benefits and Limitations of Animal Cancer Tests*, ACSH, Summit, NJ

Anon (1974). Animal models in cancer research, *Lancet*, **ii**, 1506

Baker, S. and Davey, D. (1970). The predictive value for man of toxicological tests of drugs in laboratory animals, *Br. Med. Bull.*, **26**, 208–11

Calabrese, J. (1982). *Principles of Animal Extrapolation*, John Wiley, Chichester, p. 283

Chu, K. C., Cueto, C. and Ward J. M. (1981). Factors in the evaluation of 200 National Cancer Institute carcinogen bioassays, *J. Tox. Environ. Health*, **8**, 251–80; as discussed in Salsburg (1983, p. 64)

Collingridge, D. and Reeve, C. (1986). *Science Speaks to Power*, Frances Pinter, London, pp. 39 and 47

Conning, D. (1983). Systemic toxicity due to foodstuffs. In Conning, D. and Lansdown, A. (eds), *Toxic Hazards in Foods*, Croom Helm, Beckenham, p. 11

Conning, D. (1985). New approaches to toxicity testing. In Gibson, G. and Walker, R. (eds), *Food Toxicology: Real or Imaginary Problems?*, Taylor & Francis, London

Conservation Society (1980). *Lead or Health*, Conservation Society, London

Crouch, D. (1986). Science and trans-science in radiation risk assessment, *Sci. Tot. Environ.*, **53**, 201–16

Crouch, D. (1987). The role of predictive modelling: social and scientific problems of radiation risk assessment. In Russell Jones, R. and Southwood, R. (eds), *Radiation and Health*, John Wiley, Chichester, pp. 47–63

Dalens, B., Raynaud, E. J. and Gaulme, J. (1980). Teratogenicity of valproic acid, *J. Pediatr.*, **97**, 332–3

DHSS (Department of Health and Social Security) (1980). *The Report of a DHSS Working Party on Lead in the Environment*, HMSO, London, see e.g. p. 88, para. 206

Doyal, L., Green, K., Irwin, A., Russell, D., Steward, F., Williams, R., Gee, D. and Epstein, S. S. (1983). *Cancer in Britain: The Politics of Prevention*, Pluto Press, London, pp. 84–8

Efron, E. (1984). *The Apocalyptics: How Environmental Politics Controls What We Know About Cancer*, Simon & Schuster, New York

Egger, J., Carter, C. M., Graham, P. J., Gumley, D. and Soothill, J. F. (1985). Controlled trial of oligoantigenic treatment in the hyperkinetic syndrome, *Lancet*, **i**, 540–5

Egger, J., Carter, C. M., Wilson, J., Turner, M. W. and Soothill, J. F. (1983). Is migraine food allergy? A double-blind controlled trial of oligoantigen diet treatment, *Lancet*, **ii**, 865–9

Epstein, S. (1978). *The Politics of Cancer*, Sierra Club Books, San Francisco

Gillespie, B., Eva, D. and Johnston, R. (1979). Carcinogenic risk assessment in the USA and UK: the case of aldrin/dieldrin, *Soc. Stud. Sci.*, **9**, 265–301

Gomez, M. (1981). Possible teratogenicity of valproic acid [letter], *J. Pediatr.*, **98**, 508–9

Haseman, J. K. and Salsburg, D. (1983). Letter to the Editor. *Fund. Appl. Toxicol.*, **3**, 3a–7a

Ito, N., Fukushima, S., Tsuda, H., Shikai, T., Hagiuara, A. and Imaida, K. (1985). In Gibson, G. and Walker, R. (eds), *Food Toxicology: Real or Imaginary Problems?*, Taylor & Francis, London, pp. 181–98

Kalter, H. (1968). *Teratology of the Central Nervous System*, University of Chicago Press, Chicago, p. 11; cited by Baker and Davey (1970, p. 211)

Kowlczyk, G. (1987). Interpretation of toxicity data for worker protection, *Chem. & Ind.*, 5 Oct. 1987, 681

Litchfield, J. T. (1962). Evaluation of the safety of new drugs by means of tests in animals, *Clin. Pharmacol. Ther.*, **3**, 666

Millstone, E. (1986). *Food Additives*, Penguin, Harmondsworth, chap. 4

Millstone, E. and Abraham, J. (1988). *Additives: a Guide for Everyone*, Penguin, Harmondsworth

NRPB (National Radiological Protection Board) (1984). *The Risks of Leukaemia and Other Cancers in Seascale from Radiation Exposure*, NRPB-R171, Didcot

Royal College of Physicians (1984). Joint Report of the Royal College of Physicians and the British Nutrition Foundation, on food intolerance and food aversion, *J. R. Coll. Phys. Lond.*, **18**, 115ff.

Russell Jones, R. (1984). Why radioactive doubts remain, *Guardian*, 27 Dec. 1984

Salsburg, D. (1983). The lifetime feeding study of mice and rats—an examination of its validity as a bioassay for human carcinogens, *Fund. Appl. Toxicol.*, **3**, 63–7

Salsburg, D. (1985). Personal communication, 19 April 1985

Salsburg, D. and Haseman, J. K. (1984). Letter to the Editor. *Fund. Appl. Toxicol.*, **4**, 288–92

Sune Larsson, K., Elwin, C.-E., Gabrielsson, J., Paalzow, L. and Wachtmeister, C.-A. (1982). Do teratogenicity tests serve their objectives?, *Lancet*, **ii**, 439

Supramaniam, G. and Warner, J. O. (1986). Artificial food additive intolerance in patients with angio-oedema and urticaria, *Lancet*, **ii**, 907–9

Verrett, J. and Carper, J. (1974). *Eating May Be Hazardous To Your Health*, Anchor Doubleday, Garden City, NY, chap. 5

Wilkinson, C. (1983). *Proc. 10th Int. Congress of Plant Protection*, 1983, vol.1, p. 46; quoted in *Chem. & Ind.*, 17 Dec. 1984, p. 864

Wurtman, R. J. and Maher, T. J. (1987). Effects of oral aspartame on plasma phenylalanine in humans and experimental rodents, *J. Neural. Transm.*, **70**, 169–73

5. Animal Experiments— A Failed Technology*

Robert Sharpe

'The idea, as I understand it, is that fundamental truths are revealed in laboratory experiments on lower animals and are then applied to the problems of the sick patient. Having been myself trained as a physiologist I feel in a way competent to assess such a claim. It is plain nonsense.'

Sir George Pickering, Professor of Medicine, University of Oxford. In Pickering, G. (1964). Physician and scientist, *Br. Med. J.*, **2**, 1615–19

ORIGINS OF ANIMAL RESEARCH

In 1882, during a speech to the Birmingham Philosophical Society, the great 19th-century surgeon, Lawson Tait, argued that animal experiments should be stopped 'so the energy and skill of scientific investigators should be directed into better and safer channels'. Tait undoubtedly made a major contribution towards the advance of abdominal surgery, but stated without hesitation that he had been led astray again and again by the published results of experiments on animals, so that eventually he had to discard them entirely (Tait 1882a).

Tait believed that vivisection is an error, not only because it can produce misleading results, but also because there is the constant danger that attention will be diverted from more reliable sources of information, such as clinical studies. Indeed in *Science, History and Medicine*, Cawadias (1953) argues that, whenever medicine has strayed from clinical observation, the result has been chaos, stagnation and disaster. The view needs to be taken seriously because the assumption that people and animals are alike in the way their bodies work diverted attention from the study of humans, which ultimately held back medical progress for hundreds of years. The Greek physician who was destined to dominate medicine for centuries was born in the year AD

---

*This essay was written while the author was Scientific Adviser to the British Union for the Abolition of Vivisection.

131. Galen has been described as the founder of experimental physiology and based his anatomy almost entirely on the study of apes and pigs. Galen unhesitatingly transferred his discoveries to human beings, thus perpetuating many errors (Guthrie 1945).

Unfortunately, Galen's dogmatic style, together with the Church's reluctance to allow dissection of human cadavers, meant that his errors went uncorrected for literally hundreds of years (Fraser Harris 1936). Galen's mistakes passed into current teaching and became authoritative statements of the universities until as late as the 16th century! Only with the publication of Vesalius's *Structure of the Human Body* in 1543, based on actual human dissections, did the long period of Galenic darkness begin to clear. Galen was held to blame not just for his faulty results but for using the wrong *method* (Tomkin 1973).

Yet during the 19th century, thanks essentially to just a handful of men, vivisection was transformed from an occasional, often criticised, method into the scientific fashion we know today. In 1865 the famous French physiologist, Claude Bernard, published his *Introduction to the Study of Experimental Medicine*, a work specially intended to give physicians rules and principles to guide their study of experimental medicine. Bernard regarded the laboratory as the 'true sanctuary of medical science', and considered it far more important than the clinical investigation of patients. Bernard popularised the artificial production of disease by chemical and physical means, so leading the way for today's reliance on 'animal models' in medical research. Furthermore, this influential figure created the impression that animal experiments are directly applicable to humans (Bernard 1865):

> Experiments on animals, with deleterious substances or in harmful circumstances, are very useful and entirely conclusive for the toxicity and hygiene of man. Investigations of medicinal or of toxic substances also are wholly applicable to man from the therapeutic point of view . . .

Bernard's *Introduction* was to prove the charter for 20th-century medicine but he was not alone in establishing the vivisection method. Another key figure was Louis Pasteur, whose apparently successful attempts to develop a vaccine against rabies had further glamorised the role of laboratory research and animal experiments. Pasteur's vaccine was made from the deliberately infected brain tissue of living animals but we now know that it simply does not work when injected after a rabid bite (Hattwick and Gregg 1985). Clinical experience has shown that few people bitten by a rabid animal actually develop the disease (DHSS 1977) so those who miraculously 'recovered' after a course of inoculations may not have been infected in the first place. Nevertheless Pasteur's apparent success meant that his methods were to prove highly influential, with living animals and their tissues subsequently used to produce vaccines and sera against a wide variety of conditions. This has proved a dangerous approach as contaminants from animal tissues have often produced fatal results in humans (Hayflick 1970). Furthermore the oncogenic viruses such as SV40, which contaminate tissues from primates, *only* become carcinogenic when they cross the species barrier (Hayflick 1970), so the use of human cells to prepare human viral vaccines is potentially the safest approach.

The current reliance on animal models of human disease was further popularised by the German doctor, Robert Koch, who was Pasteur's rival in developing the germ theory of disease. Koch had produced a set of rules for establishing proof that a particular germ caused the disease under investigation and one of these 'postulates' stated that, when inoculated into laboratory animals, the microbe should reproduce the same condition (Walker 1954, *Lancet* 1909). The idea was soon discredited by Koch (1884) himself during a study of cholera, when it proved impossible to reproduce the disease in animals. Ultimately Koch was forced to rely on clinical observation of patients and microscopic analysis of samples from actual cases of human cholera. As a result he was successful in isolating the microbe and discovering its mode of transmission, so that preventive action could be taken (Koch 1884). Human disease can take an entirely different form in animals so, as the *Lancet* subsequently concluded, 'We cannot rely on Koch's postulates as a decisive test of a causal organism, (*Lancet* 1909).

But occasionally Koch's animal tests seemed to work. When injected with the tuberculosis (TB) bacillus isolated from dying patients, many species—mice, guinea-pigs and monkeys but not frogs or turtles—also succumbed (Riedman 1974). Unfortunately TB takes a different form in animals (*Lancet* 1946) and human trials with Koch's highly acclaimed 'cure'—tuberculin—ended in disaster (Dowling 1977, Lehrer 1979, Westacott 1949). Nevertheless the die was cast and in the 20th century animals would be widely used as living test tubes to screen new antibacterial drugs.

ANIMAL DISEASE MODELS

Despite all of Galen's errors, and the realisation that mice are not 'miniature men', the growing influence of laboratory scientists like Bernard, Koch and Pasteur turned animal experiments into an everyday practice: medical research came to rely on artificially induced animal models of human disease. And since the direct study of human patients requires so much more skill and patience so that unnecessary risks to volunteers are avoided, it was perhaps not surprising that researchers preferred the greater convenience offered by a 'disposable species'. But with animals now being used not only to assess the value and safety of drugs, but to understand human disease, to develop surgical techniques and to acquire physiological knowledge, what are the implications for patients?

In view of the complex and often subtle nature of human disease, it is not surprising that, for the great majority of the disease entities, the animal models are considered either very poor or non-existent (Dollery 1981). Take atherosclerosis as an example. This is a condition which results in some of the West's major killers, including heart attacks and strokes, so it needs to be investigated if its aetiology is to be understood. Animal models of atherosclerosis have included birds, dogs, rats, pigs, rabbits and monkeys, but species differences have been seen in each case. The most widely used species is the rabbit, and when administered an unnatural, high-cholesterol diet, their arteries quickly become blocked, but the lesions are quite different to those found in people, in both their content and distribution (Gross 1985). While it is rare for lesions in rabbits to develop fibrosis, haemorrhage, ulceration and/or thrombosis, all of these are characteristic of lesions in human patients (Gross 1985). In animal models of stroke, it has been argued that no laboratory animal has an entirely comparable cerebrovascular supply to that of humans, and most, if not all, have a considerably greater cerebral circulatory reserve, all of which makes them far less prone to stroke (Whisnant 1958).

Animals have been used in dental research to investigate the aetiology of periodontal disease, but differences between rodents and humans could confuse clinical and epidemiological findings. To begin with, there is considerable variation in periodontal breakdown in different strains of inbred mice (Baer and Lieberman 1959). Furthermore, extrapolation from rodent periodontum to that of humans could be invalid, as both development and the potential for cementoblast differentiation in the rodent are different from those in humans (Manson and White 1983).

Huge resources have been expended on animal-based cancer research yet artificially induced cancers in animals have often proved quite different to the spontaneous tumours which arise in patients. Indeed the *Lancet* (1972) warned that, since no animal tumour is closely related to a cancer in human beings, an agent which is active in the laboratory may well prove useless clinically. This was certainly the case with the US National Cancer Institute's 25-year screening programme in which 40 000 plant species were tested for antitumour activity. As a result of the programme several materials proved sufficiently safe and effective on the basis of animal tests to be considered for clinical trials. Unfortunately all of these were either ineffective in treating human cancer or too toxic for general use (Farnsworth and Pezzuto 1984). Thus in 25 years of this extensive programme not a single antitumour agent safe and effective enough for use in patients has yet emerged, despite promising results in animal experiments. Indeed one former cancer researcher has argued that clues to practically all the chemotherapeutic agents which are of value in the treatment of human cancer were found in a clinical context rather than in animal studies (Bross 1987). Like a number of other centres, the National Cancer Institute is now using a battery of human cancer cells as a more relevant means of screening new drugs (*Scrip* 1987).

Animal cancer tests have also proved confusing in developing immunotherapy (Williams 1982). Although the techniques worked with experimental tumours in laboratory animals, and thereby raised great hopes, their clinical application proved disappointing. Once again this has been attributed to differences between the species: experimental tumours, in contrast to most human cancers, grow rapidly and are biologically different from spontaneous tumours. In addition, spontaneous cancers are less susceptible to attack by the body's defences than artificially induced cancers (Williams 1982).

Even in primates, presumably the animals closest to us in evolutionary terms, a disease can take quite a different form. The use of monkeys to investigate cerebral malaria led to the suggestion that coma in human patients is due to an increased concentration of protein in the cerebrospinal fluid, and that this leakage from the serum could be corrected with steroids (*Lancet* 1987). However, monkeys do not lapse into coma, nor do they have sequestered red cells infected with parasites, as typically seen in the human disease. In fact, steroids do not help patients and subsequent clinical investigation of the *human* condition showed that the monkey model may simply not be relevant (*Lancet* 1987).

Attempts to develop an adequate primate model of human acquired immune deficiency syndrome (AIDS) have proved equally unsuccessful. Chimpanzees are the preferred animals in AIDS research because they are the only species to maintain the virus in their bodies when injected with material from patients, although they do *not* go on to develop the disease. In 1986 Harvard's New England Regional Primate Centre reported that no deliberately infected chimpanzee had died of AIDS after more than two years under observation (King 1986):

Because of these limitations and the fact that the chimpanzee is endangered and in extremely short supply, this species will probably be of limited use to AIDS research. Attempts to infect other species of non-human primates ... with HTLV-III/LAV, have been uniformly unsuccessful.

The generally poor quality of animal models has been advanced as a strong argument for testing new drugs in volunteers and patients as early as possible to reduce the possibility of misleading predictions (Dollery 1981). Indeed it has been stated that most pharmacologists are happy if 30% of the useful actions of drugs, as determined by experiments on animals, are reproduced in humans (Saunders 1981). So it is not surprising that many of the therapeutic actions of drugs are discovered through their clinical evaluation in patients (Breckenridge 1981) or by astute analysis of accidental or deliberate poisoning (Dayan 1981) rather than by experiments on animals. This is particularly the case with many psychotropic medicines because adequate animal models of serious mental illnesses such as schizophrenia, mania, dementia and personality and behaviour disorders simply do not exist. Screening tests for the drugs generally used to treat these conditions—the major tranquillisers—are based on animal models of specific *side-effects* of existing drugs, on the assumption that many of the beneficial actions and undesirable effects of these products may be related to a similar aspect of brain chemistry, i.e. their effect on dopaminergic neurons (Worms and Lloyd 1979). Most of the rapid screening tests for potential major tranquillising drugs therefore depend on the side-effects of these drugs which are related to central dopamine-receptor blockade, for instance, catalepsy, anti-apomorphine effects and anti-emetic effects (Worms and Lloyd 1979). For example, rats are monitored for the onset of catalepsy after administration of the test drug while dogs are observed for inhibition of chemical-induced emesis. Unfortunately the nature of the tests almost inevitably means that, when successful, they lead to drugs with serious built-in side-effects such as catalepsy, Parkinsonism and tardive dyskinesia, which are also linked with dopamine-receptor blockade (Worms and Lloyd 1979). Drug-induced Parkinsonism and tardive dyskinesia have now become major problems following the treatment of serious mental illness (Melville and Johnson 1982, Stephen and Williamson 1984).

Many antidepressants were also first identified by their effects in patients. After the clinical discovery of imipramine's antidepressant properties (Sitaram and Gershon 1983), scientists accidentally found that the drug reversed chemically induced hypothermia in mice (Leonard 1984). Subsequently reserpine-induced hypothermia was used as an animal model of depression, yet several antidepressants discovered by clinical investigation were found to 'fail' the hypothermia test (Leonard 1984).

SURGERY

Clinical work has also proved the cornerstone of advances in surgery. With the rapid developments in surgical techniques following the discovery of anaesthetics in the 19th century, a number of surgeons argued strongly that advances must come from clinical practice rather than animal experiments (Beddow Bayly 1962). Tait (1882b) believed that vivisection had done far more harm than good in surgery, while the Royal Surgeon, Sir Frederick Treves (1898) issued a salutary warning about experiments he had carried out on dogs:

> ... such are the differences between the human and the canine bowel, that when I came to operate on man I found I was much hampered by my new experience, that I had everything to unlearn, and that my experiments had done little but unfit me to deal with the human intestine.

The same principles apply today when transplants and other surgical feats are being attempted. The crucial point is the underlying biological differences which make the animal experiments hazardous. It is therefore revealing that, despite thousands of experiments on animals, the first human transplants were almost always disastrous. Only after considerable clinical experience did techniques improve. At California's Stanford University, 400 heart transplants were carried out on dogs, yet the first human patients both died because of complications which had not arisen during preliminary experiments (Iben 1968). By 1980, 65% of heart transplant recipients at Stanford were still alive after one year, with the improvement attributed almost entirely to increased skill with existing anti-rejection drugs, and in the careful choice of patients for surgery (*Lancet* 1980).

In the case of combined heart and lung transplants, the early experience was once again disastrous, with none of the first three patients surviving beyond 23 days (Jamieson *et al.* 1983). In 1986 Stanford reported 28 heart–lung transplants carried out between March 1981 and August 1985. Eight patients died during or immediately after the operation. In another 10 a respiratory disorder called obliterative bronchiolitis (OB) developed after surgery, from which four more patients died and three were left functionally limited by breathlessness. The surgeons noted that (Burne *et al.* 1986):

extensive experience with animal models in this and other institutions had not indicated a serious hazard from airway disease, so the emergence of post-transplant OB as the most important complication was unexpected.

SAFETY EVALUATION

But the danger of relying on animal experiments is most vividly illustrated by the growing list of animal-tested drugs which are withdrawn or restricted because of unexpected, often fatal, side-effects in people. Examples include Eraldin, Opren, chloramphenicol, clioquinol, Flosint, Ibufenac and Zelmid (Sharpe 1988). Apart from drug withdrawal or restriction, unexpected reactions may lead to warnings from the Committee on Safety of Medicines or in the medical press (Sharpe 1988). In the case of ICI's heart drug, Eraldin, there were serious eye problems, including blindness, and there were 23 deaths. Ultimately ICI compensated more than 1000 victims (Office of Health Economics 1980). Yet animal experiments had given no warning of the dangers (Inman 1977) and even after the drug was withdrawn in 1976 the harmful effects could not be reproduced in laboratory animals (Weatherall 1982). The antibiotic chloramphenicol, passed safe after animal experiments, was later discovered to cause aplastic anaemia, which often proved fatal (Venning 1983). The *British Medical Journal* (1952) reports how the drug was thoroughly tested on animals, producing nothing worse than transient anaemia in dogs given the drug for long periods by injection, and nothing at all when given orally. Scientists have recently suggested the use of human bone-marrow cells as a more reliable means of detecting such toxic effects prior to clinical trials (Gyte and Williams 1985).

In 1982 the non-steroidal anti-inflammatory drug, Opren, was withdrawn in Britain after 3500 reports of side-effects including 61 deaths mainly through liver damage (*British Medical Journal* 1982). Prolonged tests in rhesus monkeys, in which the animals received up to seven times the maximum tolerated human dose for a year, revealed no evidence of toxicity (Dista Products Ltd 1980). Furthermore, animal tests cited in the company's literature make no mention of the photosensitive skin reactions which proved such a problem for patients (Dista Products Ltd 1980).

During the 1960s at least 3500 young asthmatics died in Britain following the use of isoprenaline aerosol inhalers (Inman 1980). Isoprenaline is a powerful asthma drug and deaths were reported in countries using a particularly concentrated form of aerosol which delivered 0.4 mg of drug per spray (Stolley 1972). Animal tests had shown that large doses increased the heart rate but not sufficiently to kill the animals. In fact cats could tolerate 175 times the dose found dangerous to asthmatics (Collins *et al.* 1969). Even after the event it proved difficult to reproduce the drug's harmful effects in animals (Carson *et al.* 1971).

Japan suffered a major epidemic of drug-induced disease in the case of clioquinol, the main ingredient of Ciba Geigy's antidiarrhoea drugs, Enterovioform and Maxaform (*Lancet* 1977a). At least 10 000 people were victims of a new disease call SMON (subacute myelo-optic neuropathy), yet animal experiments carried out by the company revealed 'no evidence that clioquinol is neurotoxic' (Hess *et al.* 1972).

Reliance on animal tests can therefore be dangerously misleading. In fact, what protection there is comes mainly from clinical trials where 95% of the drugs passed safe and effective on the basis of animal tests are rejected (*Medical World News* 1965). Nevertheless, the problem *appears* less serious than it is, because side-effects are grossly under-reported (Lesser 1980): only about a dozen of the 3500 deaths linked with isoprenaline aerosol inhalers were reported by doctors at the time, while only 11% of fatal reactions associated with the anti-inflammatory drugs, phenylbutazone and oxyphenbutazone, were reported as such (Inman 1980).

Most adverse reactions which occur in patients cannot be demonstrated, anticipated or avoided by the routine subacute and chronic toxicity experiment (Zbinden 1966). This is partly because animals simply do not have the potential to predict some of the most common or life-threatening side-effects (Welch 1967). For instance, animals cannot tell us if they are suffering from nausea, dizziness, amnesia, headache, depression and other psychological disturbances. Allergic reactions, skin lesions, some blood disorders and many central nervous system effects are some of the more serious problems which once again cannot generally be demonstrated in animals. But even when such effects are excluded, toxicity tests can still prove misleading. In 1962, the side-effects of six different drugs, reported during clinical practice, were compared with those originally seen in toxicity tests with rats and dogs (Litchfield 1962). The comparisons were restricted to those effects which animal tests have the *potential* to predict. Even so, of 78 adverse reactions seen in the patients, the majority (42) were not predicted by animal tests. In most cases, then, predictions based on animal experiments proved incorrect.

Another comparison, this time based on 45 drugs, revealed that *at best* only one out of every four side-effects predicted by animal experiments actually occurred in patients (Fletcher 1978). Even then it is not possible to tell which predictions are accurate until human trials are commenced. Furthermore the report confirmed that many common

•

side-effects cannot be predicted by animal tests at all: examples include nausea, headache, sweating, cramps, dry mouth, dizziness, and in some cases skin lesions and reduced blood pressure. But this study has an additional implication. With most of the adverse reactions predicted by animal experiments not occurring in people, there is also the danger of unnecessarily rejecting potentially valuable medicines. A classic example is penicillin which, as Florey (1953) admitted, would in all probability have been discarded had it been tested in guinea-pigs, to whom it is highly toxic (Koppanyi and Avery 1966). But the good fortune did not end there. In order to save a seriously ill patient, Fleming wanted to inject penicillin into the spine but the possible results were unknown. Florey tried the experiment with a cat but there wasn't time to wait for the results if Fleming's patient was to have a chance. Fleming's patient received his injection and improved, but Florey's cat died (BBC 1981)! Another case is digitalis. Although discovered without animal experiments, its more widespread use was delayed because tests on animals incorrectly predicted a dangerous rise in blood pressure (Beddow Bayly 1962).

One of the most common animals used in toxicity tests is the rat, yet comparisons with humans reveal major differences in skin characteristics, respiratory parameters, the location of gut flora, β-glucuronidase activity, plasma protein binding, biliary excretion, metabolism, allergic hypersensitivity and teratogenicity (Calebrese 1984). Differences in respiratory parameters (Table 5.1) are particularly important in inhalation studies, where rats are used extensively.

As 'high-risk' animal models for respiratory and cardiovascular problems, rats are considered inappropriate for asthma, bronchitis and atherosclerosis, but the species of choice for hypertension (Calebrese 1984). In fact the species most routinely used for toxicological studies are chosen not on consideration of their phylogenetic relationship to humans but on practical grounds of cost, breeding rate, litter size, ease of handling, resistance to intercurrent infections and laboratory tradition (Davies 1977).

One of the most important factors resulting in differences between the species is the speed and pattern of metabolism. Indeed reports show that variations in drug biotransformation are the rule rather than the exception (Levine 1978, Smith and Caldwell 1977, Zbinden 1963). Table 5.2 shows just how great these differences can be while Table 5.3 indicates that even with rhesus monkeys the problem is not resolved.

Toxic drug effects which are not predicted by animal tests may be seen in people when their metabolism is slower, resulting in longer exposure. But differences in the *rate* of biotransformation are only one aspect of the metabolic comparison. Of even greater importance is the *route* of metabolism. Species variability here can result in poisonous

Table 5.1   Respiratory Parameters in Rats and Humans

Respiratory parameter	Rat	Human
1. Histamine (μg g^{-1})	15.8	27.7
2. Exogenous histamine catabolism (%)	44.2	29.2
3. Histamine release (μg g^{-1}) (a) Compound 48/80 (b) Cotton dust	17.1 0	43.2 16.1
4. Lung morphometry (a) Branching angles	Decrease with increasing depth in the lung	Increase with increasing depth in the lung
(b) Symmetry	Less than humans	
(c) Diameter ratio of daughter branches at bifurcation	Greater than humans	
(d) Number of diversions of tracheobronchial tree	More variable than humans	
5. Mucous flow patterns	13.5 mm min^{-1}	15 mm min^{-1}
6. Bronchial glands	Absent	Numerous
7. Position of lung to ground	Horizontal	Vertical
8. Breathing	Obligate nose breathers	

Source: Calebrese (1984).

Table 5.2   Rates of Drug Metabolism in Various Species: Plasma Half-Lives (Hours)

Drug	Human	Rhesus monkey	Dog	Mouse	Rat	Rabbit	Cat
Hexobarbitol	6		4.3	0.3	2.3	1	
Meperidine	5.5	1.2	0.9				
Phenylbutazone	72	8	6		6	3	
Tromexan	6		21			2	
Antipyrine	12	1.8	1.7				
Digitoxin	216		14		18		60
Digoxin	44		27		9		27

Source: Levine (1978).

Table 5.3   Metabolic Half-Lives (Hours) of Various Drugs in Humans and
Rhesus Monkeys

Drug	Human	Rhesus monkey
Indomethacin	2	1.5
Isoniazid	5	5
Caffeine	3.5	2.4
Halofenate	24	16
Diazoxide	29	19
Myalex	31	3
Oxisuram	55	21
Oxyphenbutazone	72	8
Chlorphentermine	92	14

Source: Smith and Caldwell (1977).

effects which it would be impossible to predict by animal tests. A comparative study of 23 chemicals showed that in only four cases did rats and humans metabolise drugs in the same way (Smith and Caldwell 1977). One example is amphetamine, which is metabolised by the same route in humans, dogs and mice (although faster in the mouse) but by a different pathway in the rat and by still another route in the guinea-pig (Levine 1978).

These difficulties once again stress the need to assess new drugs in volunteers as early as possible. Bernard Brodie of Bethesda's National Heart Institute has stated (Brodie 1962):

> These problems highlight the importance for drug development of testing a drug in man as soon as possible to see whether its rate of metabolism makes it clinically practical. The practice of studying the physiologic disposition of a drug in man only after it is clearly the drug of choice in animals not only may prove shortsighted and time consuming, but also may result in relegating the best drug for man to the shelf forevermore.

SPECIFIC TESTS

Differences in metabolism are expected to have their greatest impact in subacute and chronic toxicity tests where animals are dosed every day for weeks, months and sometimes even years. Nevertheless the more specialised areas such as skin and eye irritancy, carcinogenicity and teratogenicity, not to mention the notorious LD50 test for systemic toxicity, have all presented problems when reliance has been placed on animal experiments. In the case of the Draize eye irritancy test, the animal most commonly used is the rabbit because it is cheap, easy to handle and has a large eye for assessing results (Ballantyne and Swanston 1977). But there are major differences which make the rabbit eye a bad model for the human eye (Ballantyne and Swanston 1977, Buehler and Newman 1964, Coulston and Serrone 1969). Unlike humans, the rabbit has a nictitating membrane and also produces tears less effectively. The acidity and buffering capacity of the aqueous humour in the eyes of human beings and rabbits are different and so is the thickness, tissue structure and biochemistry of the cornea. In humans the thickness of the cornea is 0.51 mm but it is only 0.37 mm in rabbits.

Inevitably there have been conflicting results and scientists warn that extreme caution is required in extrapolating the results from animals to the likely condition in people (Ballantyne and Swanston 1977). When the sensory irritants o-chlorobenzylidene malononitrile (CS) and dibenzoxazepine (CR) were tested in the eyes of human beings and rabbits, large differences were found, with humans being 90 times more sensitive to CR and 18 times more sensitive to CS than rabbits (Swanston 1983). On the other hand a liquid anionic surfactant formulation, with a long history as a basic ingredient of light-duty dishwashing products, caused severe eye irritation in rabbits, but extensive human experience of accidental exposure has shown it to be completely non-hazardous (Buehler and Newman 1964). The Draize test has also proved misleading in devising therapy: clinical experience in treating human eye burns caused by alkali led to the preferred treatment of thorough rinsing followed by complete denudation of the cornea. In rabbits the same technique was unsuccessful, with denudation actually *retarding* recovery threefold (Buehler and Newman 1964).

The rabbit is also the animal most frequently used to test for skin irritancy. Criticism of its predictive reliability led Britain's Huntingdon Research Centre to carry out comparative trials with six species—mice, guinea-pigs, minipigs, piglets, dogs and baboons (Davies *et al.* 1972). The researchers found 'considerable variability in irritancy response between the different species'. The most pronounced variability occurred with the more irritant materials such as an antidandruff cream shampoo where the irritancy ranged from severe in rabbits to almost non-existent in the baboon. Human volunteers suffered mild irritation. Furthermore the Huntingdon study revealed considerable differences between minipigs and piglets, which in turn differed from human responses. This is noteworthy because there appears to be a widely held belief that these species are good models for skin problems in people.

Some chemical irritants produce pain without causing structural damage and can be assessed by a variety of methods including the human blister base technique. The procedure is reported to cause little pain and produces reproducible results, with volunteers able to distinguish various intensities of discomfort (Foster and Weston 1986). Using the technique a comparison of relative potencies of three chemical irritants—o-chlorobenzylidene malononitrile (CS), n-nonanoylvanillylamine (VAN) and dibenzoxazepine (CR)— produced results which conflicted with those found from animal tests (Foster and Weston 1986). According to the blister base technique, CR is more potent than CS, which is confirmed by other human test systems, yet this is the reverse of that found from experiments with rodents. Furthermore the study found that VAN is less potent than CR, which is once again the reverse of that found from animal experiments. The authors concluded that 'data derived from humans thus appears to be of importance when assessing irritant potency'.

Experience with the Draize skin and eye irritancy tests has shown that results for the same chemical can vary widely from laboratory to laboratory and indeed within the same laboratory, because of the subjective nature of assessing the results. What is classed as a severe eye irritant by one observer may be dismissed as a mild irritant by another. This was the outcome of a study in which 25 laboratories tested 12 chemicals of known irritancy. The authors (Weil and Scala 1971) found 'extreme variation' between laboratories and concluded:

> Thus, the tests which have been used for 20 years to decide the degree of eye or skin irritation produce quite variable results among the various laboratories. To use these tests, or minor variations of them, to obtain consistency in classifying a material as an eye or skin irritant or non-irritant, therefore is not deemed practical . . . it is suggested that the rabbit eye and skin procedures currently recommended by the Federal agencies for use in delineation of irritancy of materials should not be recommended as standard procedures in any new regulations. Without careful re-education these tests result in unreliable results.

In the case of the LD50 poisoning test (LD standing for lethal dose; LD50 signifying the single dose necessary to kill 50% of test animals), results can vary widely between the species, making reliable predictions of the human lethal dose impossible. A comparison of the lethal dose of various chemicals in animals with those found from accidental or deliberate exposure in humans (Table 5.4) showed frequent extreme variations (Zbinden and Flury-Roversi 1981).

When originally introduced in 1927, the LD50 test was designed to

Table 5.4   Comparison of the Lethal Doses Found in Animal Experiments and in Humans

Sensitivity of humans compared to animal experiments		Sensitivity of humans compared to animal experiments	
Stimulants		*Chemotherapeutics*	
Pentylenetetrazol	1×	Emetine	1×
Caffeine	1×	Sulfanilamide	2–4×
Picrotoxin	1×	Quinine	6–8×
Strychnine	1×	Arsphenamine	2–30×
		Metabolic poisons	
Blood poisons		Oxalic acid	10–20×
Aniline	1×	Salicylic acid	10–20×
Potassium cyanide	1×	Arsenic	3–40×
Hydrocyanic acid	4×	Phosphorus	10–60×
Potassium chlorate	5–7×		
		Disinfectants	
		Potassium permanganate	1×
Antipyretics–analgesics		Thymol	*ca.*10×
Aspirin	1×	Mercuric chloride	3–12×
Aminopyrine	1⅓–2×	Iodoform	4–100×
Antipyrine	5–10×		
		Local anaesthetics	
		Dibucaine HCl	2–5×
Hypnotics		Tetracaine HCl	7–12×
Phenobarbital	1×	Alypin	3–30×
Tribromoethanol	2–3×	Tropacocaine	20–70×
Propallylonal	3–4×	Cocaine	4–100×
Cyclobarbital	1.5–5×	Procaine	30–150×
Carbromal	2–5×		
Diallylbarbituric acid	3–6×	*Autonomic nervous system drugs*	
Chloral hydrate	10×	Physostigmine	1×
Barbital	3–15×	Epinephrine	10–15×
Sulfonmethane	6–18×	Atropine	600–1000×
		Pilocarpine	500–2000×

Source: Zbinden and Flury-Roversi (1981).

measure the strength of drugs like digitalis, a purpose for which it has since become obsolete; but since it is easy to perform, scientists started using it as an index of toxicity for a wide variety of substances including pesticides, cosmetics, drugs, household products and industrial chemicals. And naturally the idea of a single numerical index of toxicity appealed to government bureaucrats so that the LD50 became enshrined in official government requirements for a wide range of chemical substances. According to one of Britain's largest contract houses, the Huntingdon Research Centre, approximately 90% of the LD50 tests which they perform are purely to obtain a value for various legislative needs (Heywood 1977).

It also became clear that LD50 tests could not be used to predict the results of overdose; only careful analysis of patients suffering accidental or deliberate poisoning could do that. In an account of how the National Poisons Centre at New Cross Hospital in London collates information and prepares advice on the prevention and management of drug overdosage, the Director, Dr. G. N. Volans (1986), demonstrates that 'acute toxicity data from animal tests contributes very little of value to this work'. To illustrate the failings of the LD50, Volans uses examples from two classes of drugs—the non-steroidal anti-inflammatory drugs (NSAIDs) and the antidepressants. For instance, according to Table 5.5 the safest NSAID listed appears to be aspirin, followed by ibuprofen. Yet clinical experience, Volan notes, does not accord with this since aspirin can cause death in humans at doses which are not difficult to take. On the other hand, although of supposedly greater toxicity, the largest doses of ibuprofen recorded in over 14 years of clinical use failed to produce serious toxicity, even at plasma concentrations over 20 times peak therapeutic levels. Several of the NSAIDs listed have roughly similar degrees of toxicity, yet once again this is not the case in humans because, although most appear relatively safe in overdose, it is well known that phenylbutazone can produce severe toxicity and death. Choice of a drug on the basis of its LD50 in animals could therefore be dangerously misleading.

Ultimately, preliminary tests, whatever their nature, cannot *prevent* accidental poisoning. On the other hand the introduction of child-resistant containers in 1976 resulted in a dramatic fall in hospital admissions after accidental poisoning with analgesics (Jackson *et al.* 1983).

Nor can LD50 tests be used to select dose levels of drugs suitable either for repeated administration to volunteers, or in the more

Table 5.5   LD50 Results of Some NSAIDs

Substance	Sex	LD50, rat, oral $(mg\ kg^{-1})$
Diclofenac	M	240
	F	226
Aspirin	M/F	2170
Ibuprofen	M/F	1600
Ketoprofen	–	101
Indomethacin	–	12
Naproxen	–	543
Phenylbutazone	M	608
	F	660
Mefenamic acid	–	750

Source: Volans (1986).

prolonged animal tests. This is because the toxic effects of repeated dosing cannot usually be predicted from a test like the LD50 which uses a single dose. The LD50 of dexamethasone in rats is 120 mg kg^{-1}, but on repeated administration, rats and dogs could not tolerate daily doses above 0.07 mg kg^{-1} (Zbinden and Flury-Roversi 1981). LD50 test results are also affected by genetic strain, sex, age, body weight, diseases, parasitisation, quality of feed, degree of starvation, number of animals per cage, cage size, season and climatic conditions such as temperature, humidity and air pressure (Schutz 1986). Furthermore results can be influenced by the formulation and bioavailability of test substances, quality of vehicle, administered volume, rate of application and handling during application. The LD50 can hardly be considered a biological constant and results for the same chemical can vary from laboratory to laboratory by as much as 8–14 times, using the same species and the same method of dosing (Bass *et al.* 1982).

In recent years the LD50 test has come under increasing scrutiny with a number of eminent toxicologists condemning it. Zbinden and Flury-Roversi (1981) published a major review of the LD50 and concluded that:

> For the recognition of the symptomatology of acute poisoning in man, and for the determination of the human lethal dose, the LD50 is of very little value.

In the case of carcinogenicity tests the usual problems of species variation are aggravated by the high cost and long duration (over three years) of the procedure. As a result they are unsuitable for the task of assessing the safety of more than 40 000 largely untested chemicals currently in use in our environment (Davis and Magee 1979). This must be one of the strongest arguments in favour of the quicker, cheaper and more easily reproducible *in vitro* systems which have emerged in recent years. At a conference on Public Health Control of Environmental Health Hazards held at the New York Academy of Science in 1978, Peter Hutt argued that for these very reasons reliance on animal testing for future regulatory decision-making is misplaced. Hutt (1978) was referring specifically to food additives and food safety, and observed that 'even if all animal testing facilities available in the country were deployed solely in testing the potential carcinogenicity of all food substances, it is unlikely that the project would be completed in our lifetime'. Hutt argued that the 'single most important priority for food safety policy in the future is the development, refinement, validation, and acceptance of a battery of new *in vitro* short term carcinogenicity predictive tests, on the basis of which sound regulatory decisions can be made'.

A recent report compared the results of carcinogenicity tests in the most commonly used species—rats and mice—and found that 46% of substances carcinogenic in one species were safe in the other (Di Carlo 1984). Differences in response between males and females were also found. Of 33 chemicals found carcinogenic to both rats and mice, only 13 caused cancer in both male and female animals of each species. The author concluded (Di Carlo 1984):

> It is painfully clear that carcinogenesis in the mouse cannot now be predicted from positive data obtained from the rat and vice versa.

Another study investigated whether rodent carcinogenicity tests successfully predicted the 26 substances presently thought to cause cancer in humans. An analysis of the scientific literature revealed that only 12 of these (46%) have been shown to cause cancer in rats or mice (Salsburg 1983). It was concluded that:

> . . . the lifetime feeding study in mice and rats appears to have less than a 50% probability of finding known human carcinogens. On the basis of probability theory, we would have been better off to toss a coin.

The implications for humans are obvious.

As a result of the thalidomide disaster, which left 10 000 children crippled and deformed, teratogenicity tests became a legal requirement for new medicines. While it is true that thalidomide had not been tested specifically for birth defects prior to marketing, a close analysis of the tragedy suggests that animal testing could actually have delayed warnings of thalidomide's effect on the foetus. By June 1961, Dr W. G. McBride, an obstetrician practising in Sydney, had seen three babies with unusual malformations and had strongly suspected thalidomide. To test his suspicions, McBride commenced experiments with guinea-pigs and mice but when no deformities were found, he began to have doubts that were to nag him for months (Sunday Times Insight Team 1979). Then, late in September, came further malformed babies and McBride became certain that thalidomide was responsible. He wrote to the *Lancet* and the *Medical Journal of Australia* and published his findings (McBride 1961).

Further experiments revealed that even if thalidomide had been tested in pregnant rats, the animals so often used to look for foetal damage, no malformations would have been found. The drug does not cause birth defects in rats (Koppanyi and Avery 1966) or in many other species (Lewis 1983), so the human tragedy would have occurred just the same. According to the *Catalogue of Teratogenic Agents* (Shepard 1976):

> several . . . principles were forcefully illustrated by observations made of the outbreak. The first point was that there existed extreme variability in species susceptibility to thalidomide.

The *Catalogue* reports that by 1966 there were 14 separate publications describing the effects of thalidomide on pregnant mice yet nearly all reported negative findings or else a few defects which did not resemble the characteristic effects of the drug. Only in certain strains of rabbit and primate can thalidomide's effect on the human foetus be reproduced.

Teratogenicity tests have particular problems which make the results even more difficult to extrapolate to humans than other animal tests. In addition to the usual variation in metabolism, excretion, distribution and absorption which can exist between species, there are also differences in placental structure, function and biochemistry (Panigel 1983). Foetal and placental metabolism, and the handling of foreign compounds, are different in different species, and the use of several species does not necessarily overcome the problem. The difficulties are highlighted by aspirin, a proven teratogen in rats, mice, guinea-pigs, cats, dogs and monkeys, yet despite many years of extensive use by pregnant women, it has not been linked to any kind of characteristic malformation (Mann 1984). Furthermore, if important drugs such as penicillin and streptomycin were discovered today, would we reject them because of their known ability to cause birth defects in laboratory animals (Smithells 1980)? In fact many drugs are marketed despite causing malformations in laboratory animals. British biochemist Dennis Parke (1983) gives one example:

> . . . corticosteroids are known to be teratogenic in rodents, the significance of which to man has never been fully understood, but nevertheless is assumed to be negligible. However, the practice of evaluating corticosteroid drugs in rodents still continues, and drugs which exhibit high levels of teratogenesis in rodents at doses similar to the human therapeutic dose are marketed, apparently as safe, with the manufacturer required only to state that the drug produces birth defects in experimental animals, the significance of which to man is unknown.

It is examples like this which suggest that much animal testing is more in the nature of a public relations exercise than a serious contribution to drug safety (Smithells 1980). Peter Lewis (1983), a Consultant Physician at Hammersmith Hospital in London, agrees that teratogenicity tests are 'virtually useless scientifically' but do provide 'some defence against public allegations of neglect of adequate drug testing. In other words "something" is being done, although it is not the right thing'. That 'something' is identified by D. F. Hawkins (1983), Professor of Obstetric Therapeutics at the Institute of Obstetrics and Gynaecology, and Consultant Obstetrician and Gynaecologist at Hammersmith Hospital:

> The great majority of perinatal toxicological studies seem to be intended to convey medico-legal protection to the pharmaceutical houses and political protection to the official regulatory bodies, rather than produce information that might be of value in human therapeutics.

Animal tests can also misleadingly imply advantages of a new product over existing competitors. The anti-arthritis drug, Opren, was promoted as having the potential to modify the disease process (BBC 1983), an enormous commercial advantage over existing arthritis drugs which could only alleviate symptoms at best. But Opren's apparent advantage was based on studies with artificially induced arthritis in rats (BBC 1983), which is not a good model for the human condition (*Rheumatology in Practice* 1986). Ultimately the effect could not be reproduced in humans. On the basis of animal tests another NSAID, Surgam, was promoted as giving 'gastric protection', a considerable advantage over similar drugs where gastrointestinal side-effects are a major problem. Unfortunately for the manufacturers, clinical trials showed that Surgam did indeed damage the stomach and Roussel Laboratories were found guilty of misleading advertising and fined £20 000. A report of the case in the *Lancet* stated that expert witnesses for *both* sides 'agreed that animal data could not safely be extrapolated to man' (Collier and Herxheimer 1987).

EFFECTS OF ANIMAL TESTS

Although the motivation for much testing is questionable on scientific grounds, the obsession with animal experiments has had the effect of delaying the development and introduction of safer test and monitoring systems. Yet the thalidomide tragedy should have alerted governments to the need for superior methods of safety evaluation. For instance, knowing the extreme differences in species susceptibility to the drug, the British Government could have taken a number of steps to prevent further disasters. It could have introduced legislation, as in Norway, limiting new drugs to those fulfilling a real medical need, thereby minimising potential hazards; it could have introduced really effective systems of post-marketing surveillance (PMS) for the earlier detection of side-effects; and it could have urgently promoted research into more reliable test procedures, for instance based on human cells and tissues. But the opportunities were largely neglected. Instead the Medicines Act 1968 was introduced, which simply resulted in *more* animal tests. Companies were allowed to flood the market with 'me-too' drugs, so that by 1981 the Department of Health and Social Security (DHSS) had to conclude that new chemical entities marketed in Britain during the previous 10 years 'have largely been introduced into therapeutic areas already heavily oversubscribed' (Griffin and Diggle 1981). Other surveys estimated that around 70% of new chemical entities add little or nothing to those already available (*Scrip* 1985, Steward 1978).

Little attention was given to effective PMS schemes, with the current yellow card system detecting only 1–10% of adverse reactions (Lesser 1980). Paediatrician Robert Brent (1972) has argued that efficient clinical surveillance schemes would have uncovered the link between thalidomide and limb malformations after only a handful of cases, thus preventing a major catastrophe. Another case where effective PMS would have prevented a major drug disaster is Eraldin. It took over four years to detect the drug's capacity to damage the eyes (Inman 1980). Post-marketing surveillance has recently received more serious attention with Inman's prescription–event monitoring scheme, although this is still based on a voluntary reporting system. Until recently little attempt had been made to validate human cell tests, yet researchers have argued that, despite limitations, they can give a degree of reassurance *not* provided by *in vivo* animal experiments or procedures based on animal tissues (Gyte and Williams 1985). As long ago as 1971 scientists reported that thalidomide's potential to damage the foetus could be detected in human tissue tests (Lash and Saxen 1971).

By giving so little attention to effective PMS and more reliable test systems based on human tissues, the result is that reliance is almost exclusively being placed in the original animal experiments and in clinical trials. But clinical trials only involve relatively small numbers of people so, in the absence of efficient PMS, doctors are, in effect, relying to a considerable extent on the preliminary animal experiments as a warning against adverse reactions. In fact the indications are that iatrogenic disease is reaching what DHSS Principal Medical Officer, Ronald Mann (1984) describes as 'epidemic proportions'. Government figures reveal that in 1977 over 120 000 people were discharged from or died in UK hospitals after suffering the side-effects of medicinal products (Mann 1984). Adverse reactions are considered to be an increasingly important problem with perhaps 1 in 20 general hospital beds (1 in 7 in the USA) occupied by patients suffering from their treatment (D'Arcy and Griffin 1979). In general practice as many as 40% of patients may experience side-effects (Mann 1984). One estimate puts the annual number of drug-induced deaths in Britain at 10 000–15 000 (Melville and Johnson 1982), which is 4–5 times the official estimate. Nevertheless a recent investigation of just *one* category of drugs suggests that the higher value may well be nearer the truth. The study focused on NSAIDs and estimated that they may be associated with over 4000 deaths every year in the UK from gastro-intestinal complications (Cockel 1987). Powerful chemical drugs will always carry a risk of toxic effects and the increased level of iatrogenic disease must be partly due to the rise in the number of prescriptions issued. Nevertheless reliance on tests which give so little relevant information about effects in humans, and which so often prove misleading, must *add* to the burden of iatrogenic disease. Certainly animal tests are failing to curb the current epidemic of drug-induced disease.

EFFECTS ON MEDICAL RESEARCH

The growing reliance on animal models has had a fundamental effect in medical research too, for it has often diverted attention from methods which *directly apply to humans*, such as clinical studies and epidemiology. One example is the study of diabetes. In 1788 Thomas Cawley discovered the link between diabetes and a damaged pancreas when he examined a patient who had died from the disease (Jackson and Vinik 1977). This was subsequently confirmed by further autopsies but the idea was not accepted for many years, partly because physiologists failed to induce a diabetic state in animals by artificially damaging the pancreas (Levine 1977). Eventually in 1898, Mering and Minkowski 'confirmed' the clinical findings when they produced diabetes in a dog by surgically removing the entire pancreas (Volk and Wellman 1977).

The link between smoking and lung cancer was first discovered through epidemiology and potentially represents one of the most important contributions to health policy in recent years. Yet attempts to duplicate the effect by forcing laboratory animals to inhale smoke met with little success (*Lancet* 1977b). Nevertheless negative findings in animals must have been welcome news for those who wished to deny the association.

In fact the traditional emphasis on animal-based cancer research seems to have diverted attention from a true understanding of the underlying causes, so that little attention has been given to preventive measures. Before World War I, epidemiology and clinical observation had identified several causes of cancer. It was found, for instance, that pipe smokers had an increased risk of lip cancer while radiologists often contracted skin cancer. Then in 1918 Yamagiva and Ischikawa produced cancer on a rabbit's ear by painting it with tar and the apparent potential for laboratory experiments captured the imagination of the scientific world (Doll 1980). As Sir Richard Doll (1980) explains, observational data were commonly dismissed and carried little weight in comparison with those obtained by experiment, as it was confidently believed that the mechanisms by which all cancers were caused would be discovered within a few years.

The resulting overemphasis on trying to cure the disease once it had arisen was to prove a grave mistake. A recent analysis of cancer trends in the United States indicates a substantial increase in the overall death rate since 1950, despite progress against some rare forms of the disease, amounting to 1–2% of total cancer deaths. 'The main conclusion we draw', states the report, 'is that some 35 years of intense effort focused largely on improving treatment must be judged a qualified failure'. The report concludes that 'we are losing the war against cancer' and argues for a shift in emphasis towards prevention if there is to be substantial progress (Bailar and Smith 1986). With the recent revival of epidemiology we now know much more about the major risk factors, so that 80–90% of cancers are considered potentially preventable (Doll 1977, Muir and Parkin 1985). And interestingly the United States Office of Technology Assessment report on the causes of cancer relied far more on epidemiology than on animal experiments or other laboratory studies because, its authors argued, these 'cannot provide reliable risk assessments' (Roe 1981).

The ultimate test of the success of animal experiments in medical research is whether they lead to real improvements in health which cannot be achieved by other means. There can be no doubt that life expectancy has improved considerably since the mid-19th century but this has been attributed chiefly to improvements in public health, with medical measures playing only a relatively small part (McKeown 1979, McKinlay and McKinlay 1977). By 1950 the fall in death rate had started to level out (OPCS 1978), but it was around this time that animal experiments started to increase dramatically, so has this resulted in corresponding improvements in health? In fact hospital admissions are increasing (DHSS 1976, Melville and Johnson 1982, *Annual Abstract of Statistics* 1987) as is the level of chronic sickness in all age groups (*Social Trends* 1975, 1985); more working days are being lost (Wells 1987) and the number of prescriptions issued per person has risen from an average of 4.7 in 1961 to 7.0 in 1985 (*Social Trends* 1987). The picture for major serious diseases is equally disturbing with overall cancer mortality showing no signs of decline (Smith 1982), while Britain's death rate from heart disease is one of the highest in the world (Ball 1983). Whatever animal experiments are doing, they appear to have little overall effect on our state of health.

Those who defend experiments on animals often present us with a simple choice: which life is more important, they ask, that of a child or that of a dog (Noble 1985)? Indeed the basic rationale behind animal experimentation, as spelled out by Claude Bernard (1865), is that lives can only be saved by sacrificing others. But since animal-based research is unable to combat our major health problems and, more dangerously, often diverts attention from the study of humans, the real choice is not between animals and people; rather it is between good science and bad science. In medical research animal experiments are generally bad science because they tell us about animals, usually under artificial conditions, when we really need to know about people. Only a human-based approach can accurately identify the principal causes of human disease, so that a sound basis for treatment is available and preventive action can be taken.

REFERENCES

Annual Abstract of Statistics (1987). Central Statistical Office, HMSO, London

Baer, P. N. and Lieberman, J. E. (1959). Observations on some genetic characteristics of the periodontium in three strains of inbred mice, *Oral Surg.*, **12**, 820–9

Bailar, J. C. and Smith, E. M. (1986). Progress against cancer?, *New Engl. J. Med.*, **314**, 1226–32

Ball, K. (1983). Nutrition and the public, *Lancet*, **ii**, 339–40

Ballantyne, B. and Swanston, D. W. (1977). The scope and limitations of acute eye irritation tests. In Ballantyne, B. (ed.), *Current Approaches in Toxicology*, John Wright, Bristol, pp. 139–57

Bass, R., Günzel, P., Henschler, D., König, J., Lorke, D., Neubert, D., Schütz, E., Schuppan, D. and Zbinden, G. (1982). LD50 versus acute toxicity: critical assessment of the methodology currently in use, *Arch. Toxicol.*, **51**, 183–6

BBC (1981). The discovery of penicillin, BBC Radio 4, 5 August

BBC (1983). The Opren scandal, Panorama, BBC1, 10 January

Beddow Bayly, M. (1962). *The Futility of Experiments on Living Animals*, National Anti-Vivisection Society, London

Bernard, C. (1865). *An Introduction to the Study of Experimental Medicine*; translation by Green, H. C. (1957), Dover, New York

Breckenridge, A. M. (1981). New-drug assessment in man: a clinical pharmacologist's view. In Cavalla, J. F. (ed.), *Risk–Benefit Analysis in Drug Research*, MTP, Lancaster, pp. 75–82

Brent, R. (1972). Drug testing for teratogenicity: its implications, limitations and application to man. In Klingberg, M. A., Abromovici, A. and Chemke, J. (eds), *Advances in Experimental Medicine*. vol. 27, Plenum, New York, pp. 31–43

British Medical Journal (1952). Danger of chloramphenicol, *Br. Med. J.*, **2**, 136–8

British Medical Journal (1982). Benoxaprofen, *Br. Med. J.*, **285**, 459–60

Brodie, B. (1962). Difficulties in extrapolating data on metabolism of drugs from animal to man, *Clin. Pharmacol. Ther.*, **3**, 374–80

Bross, I. D. J. (1987). Animal models: fighting cancer with a failed technology, *The Animals Agenda*, March, pp. 16–18

Buehler, E. V. and Newman, E. A. (1964). A comparison of eye irritation in monkeys and rabbits, *Toxic. Appl. Pharmacol.*, **6**, 701–10

Burne, C. M., Baldwin, J. C., Morris, A. J., Shumway, N. E., Theodore, J., Tazelaar, H. D., McGregor, C., Robin, E. D. and Jamieson, S. W. (1986). Twenty-eight cases of human heart-lung transplantation, *Lancet*, **i**, 517–19

Calebrese, E. J. (1984). Suitability of animal models for predictive toxicology: theoretical and practical considerations, *Drug Metab. Rev.*, **15**, 505–23

Carson, S., Scheimberg, J., Mackars, A. and Vogin, E. E. (1971). A continuing toxicologic reassessment of isoprenaline aerosols, *Pharmacologist*, **18**, 272

Cawadias, A. P. (1953). Corvisart and his role in the history of clinical science and the art of medicine. In Underwood, E. A. (ed.), *Science, History and Medicine*, Oxford University Press, Oxford, pp. 256–62

Cockel, R. (1987). NSAIDs—should every prescription carry a Government Health Warning?, *Gut*, **28**, 515–18

Collier, J. and Herxheimer, A. (1987). Roussell convicted of misleading promotion, *Lancet*, **i**, 113–14

Collins, J. M., McDevitt, D. G., Shanks, R. G. and Swanton, J. G. (1969). The cardiotoxicity of isoprenaline during hypoxia, *Br. J. Pharmacol.*, **36**, 35–45

Coulston, F. and Serrone, D. (1969). The comparative approach to the role of non-human primates in evaluation of drug toxicity in man: a review, *Ann. NY Acad. Sci.*, **162**, 681–706

D'Arcy, P. F. and Griffin, J. P. (1979). *Iatrogenic Diseases*, Oxford University Press, Oxford

Davies, D. M. (1977). *Textbook of Adverse Drug Reactions*, Oxford University Press, Oxford

Davies, R. E., Harper, K. H. and Kynoch, S. R. (1972). Inter-species variation in dermal reactivity, *J. Soc. Cosmet. Chem. Br. Edn*, **23**, 371–81

Davis, D. L. and Magee, B. H. (1979). Cancer and industrial chemical production, *Science*, **206**, 1356–8

Dayan, A. D. (1981). The relative worth of animal testing. In Cavalla, J. F. (ed.), *Risk–Benefit Analysis in Drug Research*, MTP, Lancaster, pp. 97–112

DHSS (Department of Health and Social Security) (1976). *Prevention and Health: Everybody's Business*, HMSO, London

DHSS (1977). *Memorandum on Rabies*, HMSO, London

Di Carlo, F. J. (1984). Carcinogenesis bioassay data: correlation by age and sex, *Drug Metab. Rev.*, **15**, 409–13

Dista Products Ltd (1980). *Opren: Clinical and Laboratory Experience*, 26 September

Doll, R. (1977). Strategy for detection of cancer hazards to men, *Nature*, **265**, 589–96

Doll, R. (1980). The epidemiology of cancer, *Cancer*, **45**, 2475–85

Dollery, C. T. (1981). Discussion session. In Cavalla, J. F. (ed.), *Risk–Benefit Analysis in Drug Research*, MTP, Lancaster, pp. 87–8

Dowling, H. F. (1977). *Fighting Infection*, Harvard University Press, Cambridge, MA

Farnsworth, N. R. and Pezzuto, J. M. (1984). Practical pharmacologic evaluation of plants, *Lord Dowding Fund Bull.*, **21**, 26–34

Fletcher, A. P. (1978). Drug safety tests and subsequent clinical experience, *Proc. R. Soc. Med.*, **71**, 693–8

Florey, H. (1953). The advance of chemotherapy by animal experiments, *Conquest*, January, Research Defence Society, London

Foster, R. W. and Weston, K. M. (1986). Chemical irritant algesia assessed using the human blister base, *Pain*, **25**, 269–78

Fraser Harris, D. F. (1936). Persecuted pioneers, *New Health*, September

Griffin, J. P. and Diggle, G. E. (1981). A survey of products licensed in the United Kingdom from 1971–1981, *Br. J. Clin. Pharmacol.*, **12**, 453–63

Gross, D. R. (1985). *Animal Models in Cardiovascular Research*, Martinus Nijhoff, The Hague

Guthrie, D. (1945). *A History of Medicine*, Nelson, London

Gyte, G. M. L. and Williams, J. R. B. (1985). The effects of some non-steroidal anti-inflammatory drugs on human granulopoiesis *in vitro*, *ATLA*, **13**, 38–47

Hattwick, M. A. and Gregg, M. B. (1985). The disease in man. In Baer, G. M. (ed.), *The Natural History of Rabies*, vol. 2, Academic, New York, pp. 281–304

Hawkins, D. F. (1983). Prescribing for pregnancy. In Hawkins, D. F. (ed.), *Drugs and Pregnancy: Human Teratogenesis and Related Problems*, Churchill Livingstone, Edinburgh, pp. 41–9

Hayflick, L. (1970). The choice of the cell substrate for human virus vaccine production, *Lab. Pract.*, **19**, 58–62

Hess, R., Keberle, H., Koella, W. P., Schmid, K. and Gelzer, J. (1972). Clioquinol: absence of neurotoxicity in laboratory animals, *Lancet*, **ii**, 424–5

Heywood, R. (1977). Supporting document in Appendix A of *The LD50 Test: Evidence for Submission to the Home Office Advisory Committee*, Prepared by Committee for the Reform of Animal Experimentation, August

Hutt, P. B. (1978). Unresolved issues in the conflict between individual freedom and government control of food safety, *Food Drug Cosmet. Law J.*, October, 558–89

Iben, A. (1968). *Erie Daily Times* (USA), 23 May

Inman, W. H. (1977). Recorded release. In Inman, W. H. and Gross, F. H. (eds), *Drug Monitoring*, Academic, New York, pp. 65–75

Inman, W. H. (1980). The United Kingdom. In Inman, W. H. (ed.), *Monitoring for Drug Safety*, MTP, Lancaster, pp. 9–47

Jackson, R. H., Craft, A. W., Lawson, G. R., Beattie, A. B. and Sibert, J. R. (1983). Changing pattern of poisoning in children, *Br. Med. J.*, **287**, 1468

Jackson, W. P. U. and Vinik, A. I. (1977). *Diabetes Mellitus*, Edward Arnold, London

Jamieson, S. W., Reitz, B. A., Oyer, P. E., Billingham, M., Modry, D.,

Baldwin, J., Stinson, E. B., Hunt, S., Theodore, J., Bieber, C. P. and Shumway, N. E. (1983). Combined heart and lung transplantation, *Lancet*, i, 1130–2

King, N. W. (1986). Simian models of acquired immunodeficiency syndrome (AIDS): a review, *Vet. Pathol.*, **23**, 345–53

Koch, R. (1884). An address on cholera and its bacillus, *Br. Med. J.*, **2**, 453–8

Koppanyi, T. and Avery, M. A. (1966). Species differences and the clinical trial of new drugs: a review, *Clin. Pharmacol. Ther.*, **7**, 250–70

Lancet (1909). Bacteriology tested by epidemiology, *Lancet*, i, 848–9

Lancet (1946). Chemotherapy of tuberculosis, *Lancet*, ii, 99

Lancet (1972). Testing anti-cancer drugs, *Lancet*, i, 827–8

Lancet (1977a). SMON and clioquinol, *Lancet*, i, 534

Lancet (1977b). Cosmetic talc powder, *Lancet*, i, 1348–9

Lancet (1980). Cardiac transplantation: the second round, *Lancet*, i, 687–8

Lancet (1987). Man over monkey, *Lancet*, i, 1016

Lash, J. W. and Saxen, L. (1971). Effect of thalidomide on human embryonic tissues, *Nature*, **232**, 634–5

Lehrer, S. (1979). *Explorers of the Body*, Doubleday, New York

Leonard, B. E. (1984). Pharmacology of new antidepressants, *Prog. Neuro-Psychopharmacol. & Biol. Psychiatr.*, **8**, 97–108

Lesser, F. (1980). Responses to treatment (review). *New Sci.*, **87**, 218

Levine, R. (1977). See the Introduction to Volk and Wellman (1977)

Levine, R. (1978). *Pharmacology: Drug Actions and Reactions*, Little, Brown, Beckenham

Lewis, P. (1983). Animal tests for teratogenicity, their relevance to clinical practice. In Hawkins, D. F. (ed.), *Drugs and Pregnancy: Human Teratogenesis and Related Problems*, Churchill Livingstone, Edinburgh, pp. 17–21

Litchfield, Jr, J. T. (1962). Evaluation of the safety of new drugs by means of tests in animals, *Clin. Pharmacol. Ther.*, **3**, 665–72

McBride, W. G. (1961). Thalidomide and congenital abnormalities, *Lancet*, ii, 1358

McKeown, T. (1979). *The Role of Medicine*, Blackwell, Oxford

McKinlay, J. B. and McKinlay, S. M. (1977). The questionable contribution of medical measures to the decline in mortality in the United States in the twentieth century, *Health and Society*, Summer, 405–28 (Millbank Memorial Fund)

Mann, R. D. (1984). *Modern Drug Use, an Enquiry on Historical Principles*, MTP, Lancaster

Manson, J. D. and White, I. M. (1983). Is periodontal disease preventable?: a review, *Proc. R. Soc. Med.*, **76**, 402–7

Medical World News (1965). Drug searchers draw on best of two worlds, *Med. Wld. News*, **16**, 46–52 and 164–72

Melville, A. and Johnson, C. (1982). *Cured to Death*, Secker & Warburg, London

Muir, C. S. and Parkin, D. M. (1985). The world cancer burden: prevent or perish, *Br. Med. J.*, **290**, 5–6

Noble, J. (1985). *Evening Telegraph* (Peterborough). 15 February

Office of Health Economics (1980). *A Question of Balance*, Office of Health Economics, London

OPCS (Office of Population Censuses and Surveys) (1978). *Trends in Mortality 1951–75*, Series DH1, no. 3, HMSO, London

Panigel, M. (1983). Monitoring human and animal intrauterine development with non-invasive methods and perfecting *in vitro* placental perfusion techniques for toxicologic and teratologic experiments. In Turner, P. (ed.), *Animals in Scientific Research: An Effective Substitute for Man?*, Macmillan, London, pp. 147–60

Parke, D. V. (1983). A more scientific approach to the safety evaluation of chemicals. In Turner, P. (ed.), *Animals in Scientific Research: An Effective Substitute for Man?*, Macmillan, London, pp. 7–28

Riedman, S. R. (1974). *The Story of Vaccination*, Bailey Bros & Swinfen, Folkestone

Rheumatology in Practice (1986). Side-effects of anti-inflammatory drugs (conference report), *Rheumatol. Pract.*, January, pp. 13–15

Roe, R. J. C. (1981). Avoidable cancer risks with specific reference to occupational factors, *Br. Med. J.*, **283**, 1421–2

Salsburg, D. (1983). The lifetime feeding study in mice and rats—an examination of its validity as a bioassay for human carcinogens, *Fund. Appl. Toxicol.*, **3**, 63–7

Saunders, L. Z. (1981). Discussion session. In Cavalla, J. F. (ed.), *Risk–Benefit Analysis in Drug Research*, MTP, Lancaster, pp. 127–8

Schutz, E. (1986). Use of acute toxicity data for pharmaceuticals. In Schuppan, D., Dayan, A. D. and Charlesworth, F. A. (eds), *The Contribution of Acute Toxicity Testing to the Evaluation of Pharmaceuticals*, Springer-Verlag, Berlin, pp. 10–19

Scrip (1985). 10 years of NCE discovery analysed, *Scrip*, 23 December, pp. 20–1

Scrip (1987). NCI's new screen for cytotoxics, *Scrip*, 16 January, p. 30

Sharpe, R. (1988). *The Cruel Deception: The Use of Animals in Medical Research*, Thorsons, Wellingborough

Shepard, T. H. (1976). *Catalogue of Teratogenic Agents*, Johns Hopkins Press, Baltimore

Sitaram, N. and Gershon, S. (1983). From animal models to clinical testing—promises and pitfalls, *Prog. Neuro-Psychopharmacol. & Biol. Psychiatr.*, **7**, 227–8

Smith, R. L. and Caldwell, J. (1977). Drug metabolism in non-human primates. In Parke, D. V. and Smith, R. L. (eds), *Drug Metabolism—From Microbe to Man*, Taylor & Francis, London, pp. 331–56

Smith, T. (1982). Action against cancer, *Br. Med. J.*, **284**, 1732

Smithells, R. W. (1980). Drug teratogenicity. In Inman, W. H. (ed.), *Monitoring for Drug Safety*, MTP, Lancaster, pp. 306–13

Social Trends (1975). Morbidity: chronic sickness 1972, Central Statistical Office, no. 6, HMSO, London, p. 134

Social Trends (1985). Chronic sickness: by sex, age, and economic activity 1982, Central Statistical Office, no. 15, HMSO, London, p. 106

Social Trends (1987). General, medical, pharmaceutical and dental services, Central Statistical Office, no. 17, HMSO, London, p. 130

Stephen, P. J. and Williamson, J. (1984). Drug-induced Parkinsonism in the elderly, *Lancet*, **ii**, 1082–3

Steward, H. F. (1978). Public policy and innovation in the drug industry. In Black, B. and Thomas, G. P. (eds), *Providing for the Health Services*, Croom Helm, London, pp. 132–53

Stolley, P. D. (1972). Why the United States was spared an epidemic of deaths due to asthma, *Am. Rev. Resp. Dis.*, **105**, 883–90

Sunday Times Insight Team (1979). *Suffer the Children—The Story of Thalidomide*, André Deutsch, London

Swanston, D. W. (1983). Eye irritancy testing. In Balls, M., Riddell, R. J. and Worden, A. N. (eds), *Animals and Alternatives in Toxicity Testing*, Academic, New York

Tait, L. (1882a). On the uselessness of vivisection upon animals as a method of scientific research. *Trans. Birmingham Phil. Soc.*, 20 April

Tait, L. (1882b). Letter, *Birmingham Daily Mail*, 21 January

Tomkin, O. (1973). *Galenism*, Cornell University Press, Ithaca, NY

Treves, F. (1898). An address on some rudiments of intestinal surgery, *Br. Med. J.*, **2**, 1385–90

Venning, G. R. (1983). Identification of adverse reaction to new drugs I: What have been the important adverse reactions since thalidomide?, *Br. Med. J.*, **286**, 199–202

Volans, G. N. (1986). Acute toxicity test data—has it any relevance for the management of acute drug overdose in man? In Schuppan, D., Dayan, A. D. and Charlesworth, F. A. (eds), *The Contribution of Acute Toxicity Testing to the Evaluation of Pharmaceuticals*, Springer-Verlag, Berlin, pp. 34–9

Volk, B. W. and Wellman, K. F. (1977). *The Diabetic Pancreas*, Tindall, London

Walker, K. (1954). *The Story of Medicine*, Hutchinson, London

Weatherall, M. (1982). An end to the search for new drugs?, *Nature*, **296**, 387–90

Weil, C. S. and Scala, R. A. (1971). Study of intra- and inter-laboratory variability in the results of rabbit eye and skin irritation tests, *Toxicol. Appl. Pharmacol.*, **17**, 276–360

Welch, A. D. (1967). Pharmacological differences, qualitative and quantitative between man and other species. In Wolstenholme, G. and Porter, R. (eds), *Drug Responses in Man*, Churchill, Edinburgh, pp. 3–23

Wells, N. (1987). A critical look at traditional measures of the economic benefits of medicines. In Teeling Smith, G. (ed.), *Costs and Benefits of Pharmaceutical Research*, Office of Health Economics, London, pp. 5–9

Westacott, E. (1949). *A Century of Vivisection and Antivivisection*, Daniel, Saffron Walden

Whisnant, J. P. (1958). Experimental cerebral vascular disease and dysfunction. In Millikan, C. H. (ed.), *Cerebral Vascular Diseases*, Grune & Stratton, New York, pp. 53–67

Williams, C. J. (1982). Immunotherapy reassessed, *Br. Med. J.*, **284**, 920–1

Worms, P. and Lloyd, K. G. (1979). Predictability and specificity of behavioural screening tests for neuroleptics, *Pharmacol. Ther.*, **5**, 445–50

Zbinden, G. (1963). Experimental and clinical aspects of drug toxicity, *Adv. Pharmacol.*, **2**, 1–112

Zbinden, G. (1966). Animal toxicity studies: a critical evaluation, *Appl. Ther.*, **8**, 128–33

Zbinden, G. and Flury-Roversi, M. (1981). Significance of the LD50 test for the toxicological evaluation of chemical substances, *Arch. Toxicol.*, **47**, 77–99

6. Trivial and Questionable Research on Animals

Clive Hollands

'Infliction of pain on an animal, then, amounts to cruelty when the pain is not compensated by the consequential good . . . The human good envisaged must be a serious and necessary good, not a frivolous or dispensable one, if the infliction of pain on animals is to be ethically acceptable.'

From the Home Office (1979) Report on the LD50 Test by the Advisory Committee on the administration of the Cruelty to Animals Act 1876, para. 12

INTRODUCTION

The new legislation governing animal experimentation in Great Britain, the Animals (Scientific Procedures) Act 1986, provides the opportunity, and in my view requires the Home Secretary, to ensure that trivial and questionable research is no longer performed on living animals in this country. It may be argued by some that such research was not in any event performed in the past under the Cruelty to Animals Act 1876. My purpose, therefore, in this essay is to provide illustrations of experiments undertaken within the past 10 years or so where the research can be considered to be either trivial, or morally or scientifically questionable, or where the degree of suffering imposed on the animals cannot be justified in relation to the purpose of the experiment.

Prince Sadruddin Aga Khan in a personal statement published in *The Observer* (1981b) referred to such work:

Medical research and scientific testing are the respectable fronts behind which our consciences shelter, but even if we accept that such ends are adequate reasons for killing and mutilating at least twenty animals a second, every day and night, all over the world, there are still the repetitive, the trivial, the needless academic exercises, the 'safety' testing of frivolous products and unnecessary drugs to be justified.

118

The full effect of Section 5(4) of the Animals (Scientific Procedures) Act 1986 has not yet, I believe, been fully realised either by those who practise vivisection or by those who oppose it:

> In determining whether and on what terms to grant a project licence, the Secretary of State shall weigh the likely adverse effects on the animals concerned against the benefit likely to accrue as a result of the programme to be specified in the licence.

This clause brings into play the 'cost–benefit' equation, which the alliance of the British Veterinary Association (BVA), the Committee for the Reform of Animal Experimentation (CRAE) and the Fund for the Replacement of Animals in Medical Experiments (FRAME) worked hard to achieve during the passage of the Bill as being the mainstay of the new legislation. The Home Office 'Guidance Notes on the Administration of the New Act' draw attention to the necessity for the Home Secretary to balance the predicted severity of the procedure against the potential benefit of the work and to set limits on the degree of severity which will be permitted. The Notes continue (Home Office 1986):

> In assessing the benefit of a project, the Home Secretary is deciding whether or not it is in the public interest (and for this purpose that includes the interests of animals), that the project should be allowed.

This clause provides the all-important 'public accountability' sought in the original CRAE paper on legislation, since the Home Secretary will in future be answerable to Parliament for all he authorises under the Act. Hence trivial and questionable research or work that imposes a degree of pain, suffering or distress on an animal which cannot be justified in relation to the purpose of the work will not be granted a project licence. As Dr Louis Goldman stated in his paper presented at the Trinity College Symposium in Cambridge (Goldman 1979):

> Experiments which cause pain and suffering demand special justification. That special justification cannot merely be that scientists are entitled to do anything that might increase the sum total of knowledge or bring about a medical advance in the remote future. There has to be a direct benefit, a good and immediate relevant reason—one which most people would accept as adequate.

That view was echoed by the British Psychological Society (1979) in the Report of the Working Party on Animal Experimentation:

To the extent that any experiment carries ethical costs, in terms of animal suffering or interference with that life pattern of animals, such an experiment requires commensurately more justification.

While it is my intention to look principally at scientific research on animals undertaken in Britain, it is nevertheless important to look also at work performed overseas, particularly in the United States, where the degree of suffering frequently inflicted on animals is unacceptable by any standards. I believe there is a moral responsibility on every scientist who uses an animal to speak out against that which is unethical, unnecessary or unacceptable whether in another country, another laboratory or nearer home.

A psychologist, Dr D. G. Boyle, writing in the *Bulletin of the British Psychological Society* some years ago, quarrelled with the use of animals in so-called 'pure' research (Boyle 1976):

> Enormous suffering is imposed on animals to establish theoretical viewpoints (possibly more in other countries than here, but the argument is universal). This is because we hold the view that the advancement of knowledge is an absolute good. I really believe that this view needs to be challenged.

Dr Boyle also emphasised that he did not believe that the amount of suffering which may be inflicted upon an animal can be left to the discretion of the experimenter.

EXPERIMENTS BROUGHT TO A HALT

The first case, therefore, that I wish to discuss illustrates this point and relates to work undertaken for a PhD thesis at St Andrews University in Scotland, which was unique for a number of reasons, not least in that it resulted in one of the few court cases against a scientist conducting animal research. It was possible to bring a prosecution under the Protection of Animals (Scotland) Act 1912 since the experiment was not covered by a licence under the Cruelty to Animals Act 1876. On 20 March 1978, Robert George Whitehead Prescott, MA PhD, was charged under Section 1 of the Protection of Animals (Scotland) Act 1912 as amended: 'That on numerous occasions with a post-graduate student, Timothy Martin Caro, he cruelly ill-treated, tortured or terrified 178 canaries, 160 laboratory mice, 17 goldfish and 2 rats in the Bute Medical Building Animal House at the University'. Timothy Caro was a post-graduate student, working under the supervision of Dr Prescott (the accused) on a thesis for his PhD. Caro originally intended to research the behaviour of the Scottish wildcat in its natural environment. Subsequently, however, after discussion with the accused, a research project into the play behaviour and prey-catching ability of the domestic feline was devised.

A special room was prepared, the floor was marked into 'scoring' squares and a number of platforms at different heights and tree branches were introduced into the room. Four groups of kittens and their mothers were used for the experiments, which, for the first six months, were concerned with play activity, using ping-pong and woollen balls as play objects. After 84 days, a second series of experiments were commenced into the prey-catching ability of the kittens, using live mice, rats, canaries and goldfish. In addition to the equipment already in the room, small pieces of tubing were used to provide escape holes for the prey. During the course of the experiments, which continued for nearly a year, between 60 and 70% of the prey animals were killed by the end of each 30-minute session. Those that escaped injury were used for further experiments, and those found injured at the end of the session were killed.

In the trial, which lasted two full days, the Home Office Inspectors in evidence considered that the experiments involved unnecessary suffering and cruelty. They stated that had an application for a licence been made, they would have recommended that the Secretary of State refuse the application. The principal witness for the prosecution was Professor Robert Hinde, Research Professor at the Department of Zoology, Cambridge, and Director of the Medical Research Council's (MRC) unit of behaviour. Professer Hinde considered the experiments cruel, and when asked if they could be considered necessary, he replied that judgement of necessary must be made on balance of the increase in scientific knowledge or alleviating human suffering against the suffering inflicted upon the experimental animals.

The Defence Advocate decided not to call witnesses and submitted that there was insufficient evidence to convict under Section 1 of the 1912 Act since the accused had not behaved cruelly, being intent only on achieving an increase in scientific knowledge. He had no cruel intention towards the animals used as prey and that such cruel intention must be proved before there could be a conviction (Hollands 1978).

The Sheriff, John C. McInnes, in giving judgement, reviewed the law and the evidence given during the trial relating to the licence and certificates held under the Cruelty to Animals Act 1876 by the accused. Based on this evidence, the Sheriff found that the experiments were not performed lawfully and the defence argument that there was no case to answer could not be accepted:

I do not think there can be any doubt that the prey used in these experiments were ill-treated, tortured, terrified and abandoned and that they were so ill-treated, tortured, terrified and abandoned as a result of being placed in a confined space with cats and/or kittens in the course of these experiments. If that had been done by a person not engaged in scientific experiments there is, in my opinion, no doubt that that would universally be regarded as cruel.

The Sheriff found Dr Prescott guilty and fined him £50, the maximum fine under the 1912 Act at that time. Sheriff McInnes emphasised in his judgement (Procurator-Fiscal 1978):

The accused is subject to the law of the land in the same way as anyone else would be when dealing with animals. There is no special protection in these circumstances for those engaged on research or experimental work in relation to acts which when done by others would be classified as amounting to an offence of cruelty. Their only protection is the 1876 Act and that protection is not available in this case.

The Secretary of State for the Home Department subsequently revoked Dr Prescott's licence under the Cruelty to Animals Act 1876. All the major national Scottish and English newspapers carried reports on each of the three days of the trial, and a number of newspapers and journals carried special features on issues raised by the trial and the vivisection controversy in general.

I have used this as a first example of questionable research, as I was closely involved with the case and had a number of meetings with the Procurator-Fiscal to discuss the issues involved, and the conduct of the prosecution both before and during the court hearing. In addition there are a number of elements which make this case important.

● Although the Home Office Inspectors indicated that, had a licence application been made to undertake these experiments, it would have been refused, nevertheless, both the *post-graduate student and his supervisor presumably considered that the planned experiments were morally and ethically justified* despite the degree of suffering to be inflicted upon the prey animals.

● During evidence given at the hearing, it was stated that Home Office Inspectors had visited the laboratories four times in 1976 and a further five times in 1977 and *at no time were they aware that these experiments were taking place.* The experiments only came to light as a result of an anonymous complaint made to the local Scottish Society for the Prevention of Cruelty to Animals Inspector. A report of the trial in the journal *Doctor* commented on the possibility that equally cruel and equally pointless experiments may be going on in other laboratories without the Home Office Inspectors being aware of them (Goldman 1978).

● An interesting footnote to the trial was that a number of zoos immediately stopped feeding live prey to their reptiles. This indicates the importance of cases coming to court from time to time which clearly has a salutary effect on all concerned.

The next example is of work which, following representations by CRAE, was referred to the Home Secretary's Advisory Committee on Animal Experiments, resulting in further experiments being stopped. This work came to my attention in 1982 as a result of a report in the now defunct *Sunday Standard* on an explosion in the rabbit population in Scotland and referred to research being undertaken on myxomatosis. According to the report, wild rabbits caught in Glenesk in Angus were injected with the myxoma virus, with the result that only 44% of them died, whereas all the domestic rabbits injected with a similar dose of the virus died.

Myxomatosis is a disease which causes the most appalling suffering, and when introduced to Britain in the 1950s was universally condemned as a method of control. The experiments which permitted wild-caught and domestic rabbits to suffer the extremely painful and distressing symptoms of the disease were undertaken, according to the Home Office (1982), to assess the resistance of local wild rabbits to myxoma virus by comparison with a known susceptible strain of domesticated rabbit. Quite clearly this is an example of so-called 'agricultural' research resulting in considerable unnecessary suffering —and for what purpose? Presumably the work was not undertaken in an attempt to find a treatment for myxomatosis. If wild rabbits have developed a natural immunity, or the virus had mutated to become less lethal, this is good news but surely such knowledge is irrelevant— unless the purpose was to consider re-introducing a more virulent form of virus into wild populations—and even a hint of such a proposal would have been universally condemned.

As a direct result of CRAE's intervention, the then Home Office Minister, Mr David Mellor MP (1984), gave me his personal view:

I share your concern about the delicate balance to be drawn between the level of suffering involved in particular experiments and the purpose of those experiments. In the circumstances the Home Secretary has decided to seek the advice of his Advisory Committee on work concerning the assessment of resistance of wild rabbits to the myxoma virus.

Subsequently further research was not authorised by the Home Office.

AGRICULTURAL RESEARCH

One of the areas of research which gives particular cause for concern under the heading of 'trivial and questionable' is that of agricultural experiments, the aim of which is merely to increase yield in one form or another, which is very different from veterinary research, studying the mechanisms of animal disease and the search for suitable treatments. However, in addition to seeking increases in yield from farm animals, there also appears to be a need for researchers to produce scientific proof of the obvious. Work undertaken over many years at the Animal Breeding Research Organisation in Edinburgh into the infant mortality of newborn lambs falls into this category. Some 95 lambs between the ages of 6 hours and 75 hours old were tested for 'resistance to body cooling in a climate chamber'. In the test, about half the lambs were clipped to leave only 2 mm of wool on neck, trunk and legs. These newborn animals were then placed in the chamber where the temperature was reduced over a period of 6 hours to *minus 20°C*. The conclusion reached in this experiment, the most recent in a series of experiments going back to 1964, was (Slee 1978):

> The experiments show that resistance of lambs to cold exposure in a climate chamber is influenced by breed, birthcoat and birth weight.

I would hazard a guess that any experienced hill sheep farmer could have provided that answer without having to subject newly born animals to such unnecessary suffering. I would also suggest that the money allocated for this research could have been better used to persuade sheep farmers to bring their ewes into lowland pastures, or at least near to the farm during the winter, and adopting the practice of indoor lambing, which would have done far more to reduce lamb mortality. The Agricultural Research Council eventually stopped further experiments, although it was denied that this was due to pressure from animal welfare groups (*Daily Telegraph* 1979). Subsequently at the same research centre, a progressively cooling water bath technique was developed to measure resistance to body cooling in newborn lambs. As reported in 1980, 429 lambs of 12 different breeds aged from half-an-hour to 62 hours old were individually totally immersed in the water except for the head, with water temperature falling from 37 to 12°C or from 25 to 10°C; and some of the lambs were tested twice. In 10% of the lambs tested, 'serious difficulty' was experienced due to the lambs being rejected by the mothers after the immersion (Slee *et al.* 1980)

Another experiment in the 'Department of the Obvious' also

undertaken in Scotland was performed at the Hannah Research Institute in Ayr. Lactating goats were starved for 48 hours to test the effects of starvation on the cardiovascular system, water balance and milk secretion (Chaiyabutr *et al.* 1980).

At least a month before the start of the experiment, each animal had one carotid artery exteriorised in a skin loop. In addition, blood vessels crossing between the two halves of the udder were ligated, and some animals had one 'milk' vein exteriorised in a skin loop. This means that the artery is brought outside the body in a loop protected by a covering of skin, making blood sampling and similar manipulations easier.

On the day before the experiment, catheters were inserted into various blood vessels and the animals were housed in a metabolism cage. Even if the animals did not suffer post-operative pain from what was quite extensive surgery, which is unlikely, they would certainly have suffered distress from being confined in a metabolic cage with various catheters connected to their bodies. Among other effects, it was found that the rate of milk secretion fell by 72% and water balance was also affected—possibly because during starvation the goats drank very little water (which was freely available). The conclusion reached was that these studies show that there are profound and relatively rapid effects of starvation in the lactating goat.

Scientists working on pig production at the Rowett Research Institute in Aberdeen were said to be 'fascinated by the challenge' of weaning piglets at or soon after birth, according to a report in the *Farmers Weekly* (Anon 1980). They were studying artificial and early weaning systems in order to relieve the sow of her lactation duties and enable her to concentrate on reproduction. A target of 30 pigs reared per sow per year looked possible.

In the test, piglets were removed from the sow at birth and placed in standard plastic flip-top kitchen bins where they remained for some three weeks being fed from automatic drinker-nipples. It would seem that there is no indignity too great that may be inflicted upon an animal if there is a chance of greater profits. The goal, it was said, is to get more out of the sow. Feed, housing and labour costs remain constant irrespective of the number of piglets she produces and rears.

Such experimental research is in direct contrast to a study of animal behaviour in a natural situation for the purpose of designing a semi-intensive husbandry system which would allow animals to indulge in a wide range of natural behaviour. An example of such work has been the development of the enriched family pig unit, which was described by Colin Tudge in an article in the *New Scientist* (Tudge 1987) as one of the most exciting recent developments in agriculture since, apart from allowing the pigs to live together in extended family groups, it was a system that was translatable into feasible commercial practice. The enriched family pig unit, developed at the Edinburgh School of Agriculture and financed initially by the Farm Animal Care Trust supported by the St Andrew Animal Fund, is a valuable step forwards in animal husbandry.

Elsewhere at other centres research is going on which is neither valuable nor essential. At the Institute of Animal Physiology in Cambridge, pigs were subjected to surgery to destroy their sense of smell and then taught to obtain food by pressing a panel switch with their snouts. According to the report of this in *Applied Animal Ethology* (Baldwin and Cooper 1979) it was found 'that their pattern of feeding was little altered by the fact that they were unable to smell their food'.

The Institute of Animal Physiology at Babraham was dubbed by the popular press some years ago as 'Frankenstein Farm' as a result of publicity given to some of the experiments performed at the centre. For example, in one experiment goats had their udders surgically removed then grafted onto their necks in order to study the flow of hormones during lactation. These experiments, it was reported, have suggested a means of detecting calves likely to die in the womb or to be born at less than normal weight and die early in life (*Observer* 1979). Apart from the immediate post-operative discomfort, these experiments were not painful in the sense that many experiments on animals cause pain. However as sometimes happens when such work is criticised, an attempt was made to suggest a medical value. In this case it was stated that quite apart from its implications for animal husbandry, such work could well be relevant to human medicine, since it is possible that similar danger signs may be detectable in pregnant women. An editorial in the *Cambridge News* (1979) on work being done at the Physiology Unit commented:

> There is a point at which one ought to admit that animals do have a basic dignity and that however important some experiments are, we accept that they trespass on their right to that dignity to an unacceptable degree.

The work described so far deals principally with agricultural research but some veterinary research can also be criticised. One example was a paper published in the *Veterinary Record* entitled 'Experimental production of infectious bovine keratoconjunctivitis' (Aikman *et al.* 1985), which reported on the experimental infection of calves with this painful condition at the University of Glasgow Veterinary School, and which resulted in the subsequent publication of letters condemning the work including one from G. M. Cooper (1985):

Am I alone in the veterinary world to feel a strong sense of revulsion at the experiment performed on those 10 calves? Was it really necessary to submit them to the pain and suffering that resulted? And has any benefit derived from their ordeal?

Much of the work described in this essay required, in addition to the Home Secretary's approval under the Cruelty to Animals Act 1876, funding from such government bodies as the Medical Research Council, the Science and Engineering Research Council or the Agricultural and Food Research Council.

TOXICOLOGY

As a direct result of the first meeting in 1977 between the group which subsequently became formalised as CRAE, and the then Home Secretary, the Rt Hon Merlyn Rees, the Advisory Committee was given the task of undertaking a review of the lethal dose 50% test, the LD50—a routine short-term toxicological study performed on animals, frequently causing considerable suffering. As the title implies, the LD50 test establishes that amount of any substance which when force-fed, inhaled by or injected into animals results in 50% of the test animals dying and 50% surviving. It is a crude and scientifically unreliable test of toxicity (Zbinden and Flury-Roversi 1981). The review of the LD50 test shows that the precision of the procedure is dependent on the number of animals used. But even with large numbers of animals there are considerable variations of the test results, because the numerical value of the LD50 is influenced by many factors, such as animal species and strain, age and sex, diet, food deprivation prior to dosing, temperature, caging, season, experimental procedures, etc. Thus the LD50 value cannot be regarded as a biological constant.

The Times (1981) in a feature on cosmetics and animal experiments referred to the 'notorious LD50 test' and CRAE's comments on the test:

> The Committee for the Reform of Animal Experimentation has produced a detailed and damning survey of LD50. They say: 'It is difficult to understand why the LD50 has gained widespread acknowledgement as a valid and predictive test . . . its use is severely challenged by a great number of scientists.'

The final report of the Advisory Committee on the LD50 test (Home Office 1979) was a disappointing document, although it did make five recommendations for reducing the severity of the test. Fortunately in the succeeding years this test has become less used, and during the passage of the new legislation on animal experiments, the Home Office Minister, Mr Mellor, gave an undertaking during Committee (Mellor 1986a):

> The LD50 test pure and simple is very rarely used in the United Kingdom nowadays. As I say, I am committed to removing regulations in the United Kingdom which needlessly require an LD50 and to use our influence in the international community to ensure that other countries mend their ways.

However, the Advisory Committee in their report did make a number of observations which are relevant to this study of trivial and questionable research (Home Office 1979):

> A cruel experiment under the Act is one where the pain caused by it was not justified by any resultant benefit or that it had been improperly conducted—as, for example, by neglect of the pain condition.
> Infliction of pain on an animal, then, amounts to cruelty when the pain is not compensated by the consequential good.
> The human good envisaged must be a serious and necessary good, not a frivolous or dispensable one, if the infliction of pain on animals is to be ethically acceptable.

Dr Perrie Adams of the Department of Psychiatry and Behavioral Sciences at the University of Texas reporting in *Scientific Perspectives on Animal Welfare* (Adams 1982) on the researcher's responsibilities in animal experimentation, argued that, in addition to teaching students why and how to answer scientific questions, we need to teach them to ask and answer other questions:

> Is this a worthwhile experiment? Is it needed, or will the results be a replication of clearly known information?
> Could the scientific problem be better answered in a non-animal model?
> Is the choice of animal appropriate?
> Is the number of animals to be used appropriate and not excessive?
> Do the procedures employed consider the animals' suffering and attempt to minimize it? In particular, are the procedures for anesthesia, analgesia, postsurgical care, and euthanasia carefully considered relative to the animals' suffering?

BEHAVIOURAL RESEARCH

Dr Adams' field of research in the behavioural sciences is perhaps one which comes in for more criticism than any other. An example of both trivial and questionable research in this field was performed in the late 1970s by Dr Arnold Chamove of the Department of Psychology at Stirling University and reported in the American journal, *Visual Impairment and Blindness* (Chamove 1978).

Eight stump-tailed macaques were separated from their mothers within one week of birth. They were reared alone in cages where they could not see other infant monkeys, although they could hear and smell them. After three months in solitary confinement, the monkeys were allowed to meet each other, but four of them did so in total darkness. The object was to assess the behaviour of the 'blind' monkeys, which it was claimed, could provide vital information on the behaviour of blind children. The project was funded by the Science Research Council out of tax-payers' money. The researchers noted that the functionally blind monkeys were almost totally lacking in aggression. They used no threats or bites. The headmaster of the Royal Blind School in Edinburgh, reported in the *Glasgow Herald* (1981), said that he was not interested in such work since the behaviour of blind children is more likely to be affected by their personal relations, the expectations of their parents and their involvement with other people. Some were aggressive because of the frustrations and pressure of blindness, while others remain completely placid.

Apart from the suffering involved in separating less-than-week-old infant monkeys from their mothers and isolating them in individual cages, the whole design of the experiments must be questioned. Abnormal behaviour in the young monkeys was more likely to have been a result of separation rather than the simulated blindness. In answer to a Parliamentary Question on these experiments, Baroness Young confirmed the Home Secretary's authorisation, details of the work done and funding, and stated (Young 1981):

This grant was for fundamental research in the field of early primate development and the respective importance of different factors influencing behaviour. Results from earlier research of this type had contributed significantly to the understanding and care of developing children.

This reply, wrapped up in important-sounding phrases, is meaningless. A letter in the *Glasgow Herald* (Tulips 1981) provided a much clearer answer:

The cruelty of the kind of experiment described in your report lies not only in the misery perpetrated on the animals but also that it should be done with such spurious justification.

The writer of the letter, Dr James Tulips, who said that he wished to express his sense of outrage at what was happening at Stirling University, was a practising psychiatrist and a former postgraduate student at Stirling. He concluded his letter by saying:

If all such experimentation ended today, psychological and human knowledge would not be significantly impoverished.

Subsequently this colony of 16 stump-tailed macaques, which were to be disposed of by Stirling University, were purchased by the International Primate Protection League. The Scottish Society for the Prevention of Vivisection, the St Andrew Animal Fund and Scottish Anti-Vivisection Society joined forces to launch the Scottish Monkey Appeal which raised the necessary money to build a new permanent home for these very disturbed animals at Edinburgh Zoo.

The catalogue of trivial and questionable research performed both in Britain and overseas in the field of psychological research is almost endless. Dr Robert Drewett of the Department of Psychology at the University of Durham, in organising a class for undergraduate students reading psychology, came across two books, both of which were available in the UK, of which he said (Drewett 1977):

Both books suggest experiments that involve cruelty of an order impossible to justify by any supposed (or real) educational value that they may have.

He then went on to describe a number of proposed experiments and concluded:

What can be said when three leading European scientists recommend for widespread class adoption, an experiment in which rats are left to swim until exhausted and on the point of drowning, in order to demonstrate that this point is reached more rapidly if weights are attached to their tails?

In the USA the situation is worse since there seems no end to the ingenuity of scientists in developing techniques to turn animals into models to study the psychological maladies of humans. At the

Primate Laboratory, Department of Psychology, University of Wisconsin, primates have been used in experiments concerned with insanity and depression. According to a report in the *International Primate Protection League Newsletter* (1981), one of the techniques used is the pit of despair, a metal box just over a metre deep, narrowing to the bottom end like a long inverted pyramid in which a monkey is totally isolated for weeks on end.

The former Managing Editor of *BBC Wildlife* magazine, David Helton, writing in that journal, made no attempt to pull his punches when discussing the work of some psychologists, of whom he said their ingenuity could strike shame into that end of the profession that tortures human beings for a living (Helton 1984).

He went on to discuss experiments into what is known as learned helplessness—experiments which are not permitted in Britain because of the degree of suffering involved. The idea here is to torment the animal into a kind of catalepsy and then present it with a way of preventing the torment. To the delight of the experimenters, it turns out that the tortured animals are less interested in preventing further torture than ones that were never tortured. This is called 'learned helplessness'. What these excited psychologists appear to overlook, Helton observed, is that it takes two to make an experiment—the monkey and the human—and the best way of judging human behaviour is not by looking at the monkey.

PRIMATE EXPERIMENTS

Another example of primate use was uncovered by Alex Pacheco of the American animal rights group, People for the Ethical Treatment of Animals (PETA). PETA was instrumental in bringing a prosecution against the researcher, Dr Edward Taub, not for the actual experiments performed, since it is almost impossible in the USA to bring such a case, but for the way in which the primates were being kept. An internationally renowned authority on primates, Dr Geza Teleki, was reported as saying that the living conditions of the animals were 'some kind of hell'. The cages were dirty, rusting, immovable and encrusted with mouldy excrement, dried blood on the walls and monkey chow lying in filthy faecal trays. There were broken wires protruding into cages. The monkeys themselves had draining wounds and limbs with stumps where there had once been fingers (Heneson 1981).

The same organisation, PETA, was involved in uncovering probably the worst example that has yet come to light of deliberate suffering and neglect of laboratory animals. The organisation was supplied with some 70 hours of video material from the University of Pennsylvania depicting in horrendous detail the suffering of many primates, mainly baboons, used in experiments into non-impact head injuries. The damage was caused by subjecting the animals' heads to acceleration forces of up to 3000*g* by means of a pneumatic ram capable of moving the head violently through an arc of 60°.

The animals were supposedly anaesthetised while the injury was inflicted, but the rapid and prolonged movement of the animals' limbs under restraint and the published research protocols (Gennarelli *et al.* 1982) clearly indicated the absence of anaesthesia. Even worse, the attitude of the researchers, which was totally unprofessional, was also distasteful in the extreme, owing to the sneering and joking which punctuated the film at the expense of the pathetic, brain-damaged caricatures of the living animals that they once were. These experiments were condemned on both sides of the Atlantic, including in this country by the then Home Office Minister, Mr David Mellor MP. The exception was the University of Glasgow, which had a direct involvement in this research in subsequently carrying out pathological examinations of some of the frozen brains of the animals. Glasgow would not condemn the work in spite of the many comments made by other scientists as well as welfarists. In a letter to the Dean of Glasgow University, reported in *The Scotsman*, Dr William McGrew of Stirling University wrote (McGrew 1985):

> The suffering portrayed, even in these fragments of tape, seemed incalculable. For someone such as I, who has studied the rich and complex behaviour of baboons in nature, to see them thus was profoundly saddening.

He continued:

> Further, the actions of the scientists were in some ways even more disturbing. It made me feel ashamed. I ask you, in all humility, would you want us as scientists to be judged by our students and by the general public on such actions?

The US Department of Agriculture (USDA) subsequently charged the University of Pennsylvania with over 70 violations of the Animal Welfare Act. According to a report in the prestigious Massachusetts Society for the Prevention of Cruelty to Animals magazine, *Animals* (1985), USDA Administrator, Burt Hawkins, stated in his report on the violations that some animals apparently were operated on without adequate anaesthesia, some were operated on under unsanitary conditions, and some were not given adequate care after they had been injured in experiments.

WEAPONS AND SAFETY TESTING

Animals are, of course, used in many other fields clearly not directly related to medicine, such as space and warfare research. In Great Britain, use of animals in warfare experiments, weapons testing or the development of weapons is not permitted. However, experiments to study wounds inflicted by weapons, providing such work is designed only to produce better methods of treatment, are permitted. There is a distinction between weapons testing and wounding experiments, but the line is a very fine one indeed. In the USA no such distinctions exist. Animals have been shot, gassed, irradiated and subjected to all the horrors of modern warfare. As long ago as 1978, dolphins were being trained to kill enemy divers since, with their built-in sonar, they were able to detect enemy divers on sabotage missions. When detected, the killer dolphins impaled them with long hypodermic needles connected to carbon dioxide cartridges—the frogmen exploded. During the Vietnamese campaign, the US Office of Naval Research is reported as stating that some 60 North Vietnamese frogmen were 'nullified' by dolphins (*Daily Express* 1977). Much more recently, dolphins trained by the US Navy at the Ocean Systems Center in San Diego have been used for 'surveillance' tasks during the Gulf War (*The Scotsman* 1987). Dolphins almost certainly are most intelligent animals and their use in warfare experiments, or for that matter for any invasive research for any other purpose, should not be tolerated in a civilised society. As Professor Meth of Seton Hall University in the United States argued in *Animals* (1977), the law is concerned with prosecuting, as criminals, persons who released experimental-subject dolphins, but the law will not ask by what right the scientists place the dolphins in captivity in the first place.

Safety testing of products in daily use in our modern sophisticated world plays an important part in all our lives. Safety testing is applicable not only to the finished product, but also to each stage of the manufacturing or preparation processes, now a legal requirement under the Health and Safety at Work Act 1974. Nevertheless, this is an area of great concern partly due to the fact that some of the tests employed (for example the LD50 test, mentioned earlier, and the Draize test, where the test substance is applied to the shaved skin or to the eyes of the test animals in order to observe irritant reactions) can cause substantial pain. Bearing in mind the degree of suffering that can be involved in such safety testing, *the question which has to be faced is whether or not this can be justified for non-essential products such as cosmetics and toiletries.*

It is true that some manufacturers have reduced the level of suffering that may be involved by using limit tests instead of the classic LD50 and terminating the tests as soon as adverse effects appear. Similarly substances which show anything more than a mild skin reaction are not normally tested in the living eye. Additionally in some testing laboratories, the superfused isolated eye taken from a dead animal is used as a pre-screen. I cannot accept in any event that such studies are justified when there are sufficient well-known ingredients that have been in use for many years, albeit originally tested on animals, which make it unnecessary to undertake further testing. The Body Shop, one of the fastest-growing retail outlets for cosmetics and toiletries, does not test any of its products on animals and requires a declaration from suppliers that none of the ingredients have been subjected to animal tests over the past five years (*Animal Testing and Cosmetics* undated). If some companies can adopt this policy, and there are quite a number who do, I see no reason why other companies cannot also do so.

Much research falls into this category—'frivolous or dispensable' (courtesy Peter Clarke and the *Guardian*

MEDICAL RESEARCH

Having dealt with a number of specific areas of research such as agriculture and psychology, I would now like to turn my attention to a more difficult area, medical research. The fact that research is medical in intent, or for that matter veterinary, is not sufficient reason to ignore limits on what may be done to living animals. Dr Harold Hewitt, who has spent a lifetime working in the field of cancer research using animals under a Home Office licence, made this point very clearly in a chapter he contributed to a book, *Animals in Research* (Hewitt 1981):

The attainment of a high standard of humanity demands that the animal experimenter consistently observes two conditions; firstly, he must deny himself knowledge, however valuable it may appear, which cannot conceivably be obtained without the infliction of suffering; secondly, he must be prepared to devote a large part of his diligence and ingenuity to the design of procedures and end-points which enable the information he seeks to be gained before the onset of distress resulting from interference. To refuse the first condition is arrogant; to fail in meeting the second may often betray a want of research ability.

An example of work which went beyond those bounds was performed at the Western Infirmary in Glasgow in a study of prostaglandins. Scientists stressed a group of rats following a two-day treatment with a drug which inhibited production of prostaglandins. The 'necessary' stress was induced, under a short-lasting anaesthetic, by fracturing the right femur of the animals using dental forceps. In a private communication, Dr Hewitt (1980) told me that he would not have accepted this paper as a referee and would have advised an editor not to publish it on the grounds that the animals had been subjected to 'unacceptable levels of pain' since they had been confined to metabolism cages, given a drug which in the higher doses gave intestinal ulceration, perforation and peritonitis, had one femur crushed and fractured by dental pliers under a very short-lasting anaesthetic and then had repeated cardiac punctures; into the bargain they were fed a pretty dismal synthetic diet. Dr Hewitt added that such work was deplorable and is the kind of thing he would prohibit.

Earlier in this essay I discussed the maternal deprivation experiments performed at Stirling University using stump-tailed macaques which comprised a study into the effects of blindness. Work at the other end of the spectrum is another popular area of study—vision research. One of the earliest examples of such work I have on record dates back to 1972 and was reported in great detail in the *New Scientist* (Humphrey 1972) under the title 'Seeing and nothingness'. The study involved a monkey called Helen who had the visual cortex of the brain surgically removed, as a result of which she had been able to recognise nothing. Immediately after the operation, she appeared quite blind and during the next year she showed little sign of having any vision. The surgery did not actually damage the eyes but the part of the brain which interprets visual messages received from the eyes. Publication of the article drew an immediate and angry response from readers of the *New Scientist* which continued for over a month until the editor closed the correspondence. It was perhaps the attitude of Dr Humphrey towards his animal subject rather than the experiment itself which caused much of the concern. For example, at one point he reported that at one stage the work was interrupted when he moved from Cambridge to Oxford. Helen also moved, but as Dr Humphrey had a thesis to finish, Helen was left to her own devices for about 10 months—such devices, that is, as she could manage in a small cage.

It is worth noting that later Dr Humphrey (1976) admitted:

Some years ago I made a discovery which brought home to me dramatically the fact that, even for an experimental psychologist, a cage is a bad place in which to keep a monkey. Since that time, in working with laboratory monkeys I have been mindful of the possible damage that may have been done to them by their impoverished living conditions.

Another scientist working in the field of vision research, Professor Blakemore, has been involved for many years at both Cambridge and Oxford Universities in experiments such as keeping kittens in environments where the only things they could see were walls painted in vertical or horizontal stripes. After some six months he found that, unsurprisingly, the kittens could only see, when placed in a normal visual environment, those objects painted with either vertical or horizontal stripes. Since that time Professor Blakemore has moved on and has been involved in many other forms of experimental research into vision in which he has, for example, stitched up one or both eyes of kittens and monkeys to judge the effect, over a period of time of up to 5 years, on the animals' ability to see (Swindale *et al.* 1981, Price and Blakemore 1985).

Dr Louis Goldman, in the magazine *Doctor* (Goldman 1977), commenting on such experiments, said that, although he was not a neurophysiologist and could offer no informed opinion about the scientific worth of the studies, he would question the cost in animal suffering needed to obtain the information even if it were regarded as important by the scientific community. As a physician with some experience of scientific investigation, reading the descriptions of the experiments aroused an intense feeling of distaste and repugnance, he added.

At the time of writing, these experiments continue despite vocal public concern.

A further example of work involving the eyes, which can only be described as questionable at best, was performed at the Institute of Neurology in London. Young kittens were subjected to surgery under anaesthesia to remove muscles and tissues from the eyes as well as the nictitating membrane (the third eyelid). These mutilated animals were then trained to lick a fish reward while being shown a scene which was to represent a 'safe' stimulus and to stop licking if a 'dangerous' stimulus was presented. Failure to respond resulted in the kittens being given an electric shock as punishment (Jacobson and Ikeda 1979).

In the field of medical research and, indeed, in every other area of research, an issue which I would place under the heading of questionable has nothing to do with the experiment itself, but with the lack of expertise and experience of those performing the actual scientific procedures.

The *Veterinary Record* (Dutoit *et al.* 1981) published a report of experiments performed at the Nuffield Department of Surgery at the University of Oxford, in which 138 mongrel dogs received renal transplants and 30 puppies, also presumably mongrels, received pancreatic fragments. These experiments at a premier university were cause for concern in view of the suffering endured by at least 10 of the adults and five of the puppies which developed intussusception. Symptoms of this condition, which is a prolapse of one part of the intestine, are vomiting, attacks of pain, dehydration, failure to eat, weight loss and straining with the passage of blood-stained mucus. While the actual transplantation of the kidneys would have been performed under total anaesthesia, this is an example of the suffering which can follow such major interference. Correspondence which followed publication of the original article commented unfavourably on the general state of the animals' health before surgery (untreated roundworm infection, respiratory tract infections and diarrhoea) and the lack of knowledge displayed by the researchers in not preventing the prolapse. The writers, vets (Clayton-Jones and Gerring 1981), asked whether it is too much to expect that those who are best trained in the welfare, treatment and diagnosis of animal diseases should be consulted when animals are to be subjected to medical experimental surgery in order that such wasteful and distressing complications can be minimised.

Hopefully the requirement under new legislation for every research laboratory to appoint a named veterinary surgeon, with a statutory responsibility for the welfare and well-being of all animals in the laboratory, will prevent such unnecessary suffering.

OTHER CAUSES FOR CONCERN

Another issue which does not relate to the actual scientific procedures performed on the animal, or for that matter, the purpose of the work, is the question of housing and facilities for exercise, play and social contact, particularly for larger animals—primates, dogs, cats and farm animals—used for scientific purposes. Irrespective of the scientific procedures involved, the lack of these basic facilities, particularly in long-term studies, is possibly of greater concern than the actual procedures to which the animal will be subjected. Cyril Rosen, the UK representative of the International Primate Protection League, summed this up by saying (Rosen 1985):

> For an active animal to have no means of exercising its limbs and for an intellegent animal to have no provision for mental occupation is an act of cruelty.

Dr Andrew Rowan, Assistant Dean at Tufts University School of Veterinary Medicine in the USA, had this to say on the subject of primate housing in his book *Of Mice, Models and Men* (Rowan 1984):

> Primates are generally housed in laboratories under conditions that are not just impoverished, there is almost no environmental enrichment. There have been a few initiatives to change this, but far too many primates spend their lives (a chimpanzee can live for fifty years in captivity) in barren cages that provide little more than room to turn around and stretch.

The Royal College of Surgeons in undertaking long-term dental studies using primates, and the Institute of Psychiatry in inducing epileptic fits in baboons by subjecting them to strobe lighting effects, were guilty in this regard. Leaving aside any question of the justification of the work itself, in both instances non-human primates were kept over a period of years in standard primate caging with no facilities for exercise, social contact or play. I am not prepared to accept that treating animals in this way can ever be justified for whatever purpose. I hope that the provision in the new legislation requiring the severity grading to include the way animals are housed and what facilities are provided for exercise and play, will go some way to righting this situation. Laboratory animals should be provided with that which they require to enable them to indulge in their natural behavioural patterns irrespective of the cost to the laboratory.

In this survey of research undertaken in Britain and overseas which falls into the category of trivial or questionable, I should perhaps include another category 'bizarre'. For example, a wildlife biologist in the USA apparently conducted research which proved that the survival rates of crippled ducks in the wild were much lower than many people suppose. He took the trouble to capture 135 mallards, broke the wings of 74 of them and bound the wings of the rest with a leather strap.

He then proceeded to release half in a marsh and kept the others in a pen to see what would happen (*The Unicorn* 1983). At Exeter University researchers were surprised to find that chickens developed the ability to recognise photographs of other hens. The importance of this work, the scientist involved is reported as saying, was that if we could learn to understand how birds discriminate between things, it might help the deaf and dumb (*Mail on Sunday* 1983). While in Australia, according to a report in *The Times* (1983), scientists believe that within five years, using genetic engineering techniques, it will be possible to produce a sheep which, when fed a chemical substance, would shear itself.

The field of genetic engineering is one which, in my opinion, provides the same kind of possibilities for good or evil as splitting the atom did many years ago. By using these techniques it may be feasible to develop life-saving drugs more easily and by safer methods than at present. It is also possible to produce genetically novel animals, and in America such creatures can be patented. Farm animals could be created for example, which could provide meat in the right places on their bodies to give the cuts that are most popular—irrespective of the cost to the animal. Other possibilities are even more horrifying. It was reported in *The Daily Telegraph* (1980) with some scepticism that a scientist in the People's Republic of China had successfully impregnated a chimpanzee with human sperm but that the animal died during the third month of pregnancy.

More recently, Dr Geoffrey Bourne of the world-famous Yerkes Primate Center in the USA, commenting on the proposal that American scientists were ready to inseminate artificially a female ape with human sperm, was reported as saying (*Observer* 1981a) that it was not just a wild idea. There would be quite a good chance of getting the chimpanzee pregnant. He wanted to go ahead with the experiment but the Primate Center decided against it because of fears that media coverage would have led to funds being cut off by the US Government.

And in 1987 an Italian scientist, Professor Chiarelli, revealed that he had been involved with a successful cross-fertilisation achieved in the USA but due to 'moral, social and ethical considerations' the half-human, half-ape creature was killed before actual birth (*London Evening Standard* 1987). He added that developments in biogenetic mutations have made the operation possible.

The Daily Telegraph's Peking correspondent reported that the original work in China came to light in a Shanghai newspaper (*Wen Hui Bao*) which quoted the scientist as saying that;

> The half-man, half-ape result of crossing a chimp with a human would be able to do productive work. It would be creative and

intelligent enough to learn simple words, and could become a source for organ transplants.

The possibilities opened up by genetic manipulation make the fantasy of such 'mad scientist' dreams that much closer to reality. We could create a new slave race, a kind of stone-age zombie. My only concern over such 'bizarre' experiments is the suffering inflicted on animals. If animals were not involved, research of this kind would depend on the conscience of the scientist and on the public who largely finance research through taxes.

In concluding this chapter, I want to look to the future. When the new legislation on animal experimentation was going through Parliament in 1986, the Home Office Minister responsible, Mr David Mellor MP, stated quite clearly (Mellor 1986c):

> The reduction in the number of animals used and the reduction of suffering is at the heart of the Bill.

He also expressed his hopes for the new legislation (Mellor 1986b):

> We are trying to ensure—and I believe we shall succeed in that ambition; certainly we give ourselves the tools to stand a better chance of succeeding in that ambition within the Bill than we have available at the moment—that no animal is needlessly used in a laboratory and that, when animals have to be used, they are used in accordance with the strictest conditions that a civilised country can devise.
>
> I stress that even when animals have to be used they are used in accordance with the highest standards, so the animal welfare element predominates over everything.

There is no way that we can end all animal-based research overnight. But we can ensure, and hopefully the new legislation *will* ensure, that the kind of trivial, questionable and bizarre work discussed in this essay will not in future be permitted. As I mentioned in the opening paragraphs, the cost–benefit equation which links not only the pain, suffering and distress inflicted on the animal, but also any other interference including facilities for housing, exercise and play, with the purpose of the project, will play a valuable part in ensuring that work of an ethically or scientifically dubious nature is not permitted. Professor Patrick Bateson FRS, Secretary of the Ethical Committee of the Association for the Study of Animal Behaviour, used an excellent diagrammatic illustration of the cost–benefit equation in an article

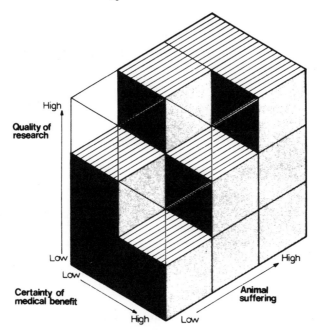

Professor Patrick Bateson's excellent demonstration of the *cost–benefit equation*: 'When a research proposal falls into the opaque part of the cube, the experimental work should not be done'. (This first appeared in *New Scientist*, the weekly review of science and technology)

published in the *New Scientist* entitled 'When to experiment on animals' (Bateson 1986). This illustration demonstrates more effectively than words how the system must work in future.

But finally, the responsibility devolves on each one of us, whether animal welfarist, or scientist, or politician, to speak out against that which is unethical, unnecessary or unacceptable, whether performed overseas, or in another laboratory, or nearer home.

REFERENCES

Adams, P. (1982). In Dodds, W. J. and Orlans, F. B. (eds), *Scientific Perspectives on Animal Welfare*, Academic, New York, p. 41

Aikman J. G., Allan, E. M. and Selman, I. E. (1985). Experimental production of infectious bovine keratoconjunctivitis, *Vet. Rec.*, **117**, 234–9

Animal Testing and Cosmetics (undated). Body Shop, Sussex, leaflet

Animals (1977). Massachusetts Society for the Prevention of Cruelty to Animals, November/December

Animals (1985). Massachusetts Society for the Prevention of Cruelty to Animals, December

Anon (1980). Research—Rowett seeks earliest possible weaning date. *Farmers Weekly extra—Pig Production*, 11 July

Baldwin, E. A. and Cooper, T. R. (1979). The effects of olfactory bulbectomy on feeding behaviour in pigs, *Appl. Anim. Ethol.*, **5**, 153–9

Bateson, P. G. (1986). When to experiment on animals, *New Sci.*, **109**, 30–2

Boyle, D. G. (1976). Animals as experimental subjects (letter), *Bull. Br. Psychol. Soc.*, **29**, 312

British Psychological Society: Scientific Affairs Board (1979). Report of a Working Party on Animal Experimentation, *Bull. Br. Psychol. Soc.*, **32**, 44–52

Cambridge News (1979). Cambridge, 10 July

Chaiyabutr, N., Faulkner, A. and Peaker, M. (1980). Effects of starvation on the cardiovascular system, water balance and milk secretion in lactating goats, *Res. Vet. Sci.*, **28**, 291–5

Chamove, A. S. (1978). Deprivation of vision in social interaction in monkeys, *Vis. Impairment Blindness*, **72**, 103

Clayton-Jones, D. G. and Gerring, E. E. L. (1981). Renal transplant surgery in the dog (letter), *Vet. Rec.*, **108**, 83

Cooper, G. M. (1985). Infectious bovine keratoconjunctivitis (letter), *Vet. Rec.*, **117**, 394

Daily Express (1977). London, 1 December

Daily Telegraph (1979). London, 22 November

Daily Telegraph (1980). London, 13 December

Drewett, R. F. (1977). On the teaching of vivisection, *New Sci.*, **76**, 292

Dutoit, D. F., Homan, W. P., Reece-Smith, H., McShane, P., French, M. E., Denton, T. G. and Morris, P. J. (1981). Canine intestinal intussusception following renal and pancreatic transplantation, *Vet. Rec.*, **108**, 34–5

Gennarelli, T. A., Thibault, N. E., Adams, J. H., Graham, D. I., Thompson, C. J. and Marcinin, R. P. (1982). Diffuse axonal injury and traumatic coma in the primate, *Ann. Neurol.*, **12**, 564–74

Glasgow Herald (1981). Glasgow, 23 February

Goldman, L. (1977). Animal machines, *Doctor*, **7**, 18–19

Goldman, L. (1978). Controversy: experiments on animals, *Doctor*, **8**, 15

Goldman, L. (1979). In Paterson, D. A. and Ryder, R. (eds), *Animal Rights—A Symposium*, Centaur, Fontwell, p. 191

Helton, D. (1984). *BBC Wildlife*, London, August

Heneson, N. (1981). Cruelty to animals: the state versus the scientist, *New Sci.*, **92**, 672–4

Hewitt, H. (1980). Personal communication

Hewitt, H. (1981). In Sperlinger, D. (ed.), *Animals in Research—New Perspectives in Animal Experimentation*, John Wiley, Chichester, pp. 170–1

Hollands, C. (1978). Personal notes taken at the trial Cupar *v*. Prescott, at Cupar Sheriff's Court, 20–22 March

Home Office (1979). Advisory Committee Report on the Enquiry into the LD50 Test, para. 12

Home Office (1982). Letter to the Committee for the Reform of Animal Experimentation, 23 July

Home Office (1986). *Guidance on the Operation of the Animals (Scientific Procedures) Act 1986*, para. 50

Humphrey, N. (1972). Seeing and nothingness, *New Sci.*, **53**, 682
Humphrey, N. (1976). In Bateson, P. P. G. and Hinde, R. A. (eds), *Growing Points in Ethology*, Cambridge University Press, Cambridge, pp. 303–17
International Primate Protection League Newsletter (1981). Monkey depression experiments at University of Wisconsin, *Int. Primate Prot. League Newsl.*, **8**, March, 8–9
Jacobson, S. G. and Ikeda, H. (1979). Behavioral studies of spatial vision in cats reared with convergent squint—is amblyopia due to arrest of development? *Exp. Brain Res.*, **34**, 11–26
London Evening Standard (1987). London, 11 May
Mail on Sunday (1983). London, 24 July
McGrew, W. (1985). *The Scotsman*, Edinburgh, 17 June
Mellor, D. (1984). Letter to Committee for the Reform of Animal Experimentation, 24 October
Mellor, D. (1986a). *Hansard House of Commons Reports*, London, Standing Committee A, col. 134, 11 March
Mellor, D. (1986b). *Hansard House of Commons Reports*, London, Standing Committee A, col. 541–2, 20 March
Mellor, D. (1986c). *Hansard House of Commons Reports*, London, 96(97), 21 April
Observer, The (1979). London, 18 November
Observer, The (1981a). London, 1 March
Observer, The (1981b). London, 16 August
Price, D. J. and Blakemore, C. (1985). The postnatal development of the association projection from visual cortical area 17 to area 18 in the cat. *J. Neurosci.*, **5**, 2443–52
Procurator-Fiscal (1978). Cupar *v.* Prescott, Written Judgment
Rosen, C. (1985). Letter to the Universities Federation for Animal Welfare, 1 March
Rowan, A. N. (1984). *Of Mice, Models and Men—A Critical Evaluation of Animal Research*, State University of New York Press, Albany, NY, p. 113
Scotsman, The (1987). Edinburgh, 28 October
Slee, J. (1978). The effects of breed, birthcoat and body weight on the cold resistance of newborn lambs, *Anim. Product.*, **27**, 43–9
Slee, J., Griffiths, R. G. and Samson, D. E. (1980). Hypothermia in newborn lambs induced by experimental immersion in a water bath and by natural exposure outdoors, *Res. Vet. Sci.*, **28**, 275–80
Swindale, N. V., Vital-Durand, F. and Blakemore, C. (1981). Recovery from monocular deprivation in the monkey III: Reversal of anatomical effects in the visual cortex. *Proc. R. Soc. Lond. (Biol.)*, **213**, 435–50
Times, The (1981). London, 28 October
Times, The (1983). London, 7 September
Tudge, C. (1987). Agriculture: time to press ahead, *New Sci.*, **115**, 40–7
Tulips, J. (1981). *Glasgow Herald*, Glasgow, 28 February
Unicorn, The (1983). Elkins, PA, vol. 3(3), March
Young, Baroness (1981). *Hansard House of Lords Parliamentary Reports*, 26 February
Zbinden, G. and Flury-Roversi, M. (1981). Significance of the LD50 test for the toxicological evaluation of chemical substances, *Arch. Toxicol.*, **47**, 77–99

7. Replacing Animal Experiments

Martin Stephens

'Those who experiment upon animals by surgery and drugs, or inoculate them with diseases in order to help mankind by the results obtained, should never quiet their consciences with the conviction that their cruel action may in general have a worthy purpose. In every single instance they must consider whether it is really necessary to demand of an animal this sacrifice for men.'

Albert Schweitzer (1923). In Joy, C. R. (ed.) (1950). *The Animal World of Albert Schweitzer*, Beacon, Boston, p. 190

INTRODUCTION

People seeking change on behalf of animals in laboratories can be broadly categorised as abolitionists or reformists. Abolitionists seek to ban all animal experiments, or at least those experiments that are not in the best interests of the animal subjects. Reformists, or welfarists, seek better safeguards for animals subjected to experimentation. Spanning both categories are people seeking to transform the research process so that animal-based methods are replaced by non-animal methods.

Russell and Burch (1959) provided a framework for this transformation. These British scientists advocated efforts to develop research methods that could *replace* the use of animals in laboratory procedures. Russell and Burch also called for the development of interim measures that, while animals were still being used, could *reduce* the numbers used or *refine* such use so that less pain or suffering ensued. Replacement, reduction and refinement constitute the three 'R's of the 'alternative approach' to laboratory practices. The ultimate goal of this approach is the complete replacement of laboratory animals with non-animal methods that are at least as scientifically sound (some would say unsound) as animal-based methods.

Complete replacement is an idealistic goal, one likely to take decades to attain. This realisation should not dissuade us from taking the pursuit seriously nor serve to discredit the potential of the alternatives approach. Even interim steps towards the goal of complete replacement have the potential to spare millions of lives.

The alternatives approach is supported to some extent by virtually all groups along the spectrum of the vivisection controversy. A small but increasing number of scientists see the search for alternatives as a way to depolarise this controversy. Alarmed by thefts of research animals and damage to laboratories by the Animal Liberation Front and similar groups, they are eager to establish a common ground with more moderate critics.

Reduction and refinement are discussed elsewhere in this volume. The present essay is devoted to replacements, specifically their nature, historical and current applications, and future prospects, as well as incentives to pursue replacements, sources of funding for research on replacements, and scientists' attitudes towards the alternatives approach.

TYPES OF REPLACEMENT

Replacement alternatives are methods that can eliminate the 'need' for animal subjects in particular experiments. Employing replacement methods does not necessarily mean that animal subjects are not involved in any phase of a research or testing programme. For example, replacement methods may substitute for animals in preliminary screening of drugs for efficacy or toxicity, yet animals may still be involved in testing those drugs that show promise in the screen. Employing various replacement techniques in combination, rather than in isolation, should facilitate the attainment of complete elimination.

Replacement methods span a wide variety of procedures and systems, including *in vitro* methods, computer modelling, studies of organisms of limited or no sentience, physical and chemical methods, human studies and several miscellaneous methods.

In vitro methods

In vitro methods enable researchers to study a wide array of body components under carefully controlled conditions outside the body. *In vitro* systems can contain subcellular components, isolated cells, tissues derived from biopsies, tissue slices from whole organs, or whole organs treated with special chemicals. *In vitro* literally means 'in glass', a reference to the test tubes or other apparatus containing the samples under study. The term contrasts with *in vivo*, which refers to studies of intact organisms.

In vitro systems are a prominent part of current research into alternatives, especially those systems that involve the culturing of cells (cell culture), tissues (tissue culture) and organs or organ fragments (organ culture) in a nutrient medium. The term 'tissue culture' often refers to all three culture systems. In cell culture, a tissue fragment is dissociated into its component cells. The first generation of these cells is a 'primary cell culture'. If the cells multiply indefinitely, a 'continuous cell line' is established. In tissue culture (in the strict sense), tissue fragments are not dissociated into component cells. In organ culture, the aim is to maintain the three-dimensional structure and function of organs or organ fragments. Organ cultures are relatively short-lived and do not propagate themselves, so fresh samples are needed each time cultures are initiated. *In vitro* technology can be applied to study virtually any cell type in the body. The practical problem of growing specialised cells has now been largely solved (Griffiths 1984).

In vitro techniques have several advantages over *in vivo* techniques. They enable cells, tissues or other components to be studied apart from the potentially confounding influences of other body systems. Because chemicals of interest can be added directly to the culture, instead of first passing through the digestive and circulatory systems, much smaller amounts of chemicals are needed. This sensitivity was the main reason why the US National Cancer Institute (NCI) recently launched a $2.5 million screening programme for antitumour agents. An NCI representative noted that 'the materials that we are typically looking for are trace constituents, so the *in vivo* model is inherently an insensitive one and we may miss, in most cases, our most interesting lead' (*Blue Sheet* 1985). Another advantage of *in vitro* systems is that the cells to be cultured can be cloned to achieve greater homogeneity or be manipulated in other desired ways, and then studied.

The main advantage of *in vitro* systems—controlled study of an isolated system—is also a disadvantage. What occurs in isolation may not occur when the complex systems of the body interact, and hence *in vitro* methods by themselves are not likely fully to replace *in vivo* studies. Nevertheless, studying individual components of complex systems is a well-known paradigm of scientific investigation. Moreover, *in vitro* systems can be modified to reflect interactions within the body better. For example, tissue derived from one organ can be exposed to specific hormones produced elsewhere in the body, or potentially toxic chemicals can be incubated with liver cells to determine if the liver detoxifies them before they can exert their toxic effects.

From an animal-welfare perspective, the main advantage of substituting *in vitro* systems for *in vivo* systems is that animals are no longer employed as experimental subjects. Animals may still be used as sources of cells. However, this material can be derived from minimally invasive biopsies or from slaughterhouse by-products. Even if animals are killed specifically for their cells, many cultures can be derived from the tissues of a single animal. And those cells, once cultured, can be grown and harvested for as long as they retain their essential characteristics. Of course, animals are not the only sources of tissue for *in vitro* studies. Human tissue can be derived from autopsies, biopsies and other sources. The study of human tissue obviates the need to extrapolate findings from animal models to the human condition.

In vitro technology has had a considerable impact on biomedical science. According to a recent report on alternatives by the US Congress's Office of Technology Assessment (OTA 1986), which was otherwise fairly conservative and defensive of the *status quo*, 'There is virtually no field of biomedical research that has not been affected by *in vitro* technology'.

A few applications of *in vitro* methods will be noted briefly here.

● Tissue culture techniques can be applied in screening substances for their potential as antiviral drugs. For example, one pharmaceutical company that began using mice to screen for antiviral drugs later added a cell culture as a primary screen and organ culture as a secondary screen and retained mice as a final screen. In 1963, the company screened 1000 substances per year using approximately 16 000 mice. Twelve years later, after adopting *in vitro* techniques, it screened 22 times more substances per year using approximately one-tenth the number of mice (Rowan 1979).

● Tissue culture has been applied with great success in finding alternatives to animals in the production of vaccines. The first vaccines produced from tissue culture were the Salk and Sabin polio vaccines, introduced in 1953 and 1956, respectively. Since then, many other vaccines have been produced in this manner. Indeed, according to Smyth (1978, p. 81), author of *Alternatives to Animal Experiments*, 'vaccine production is the way in which tissue culture can most effectively be used as an alternative to living animals'.

● An *in vitro* test of the potency of yellow fever vaccine has been approved by the British Licensing Authority as a replacement for an *in vivo* test (Stratmann *et al.* 1987). A continuous cell line derived from monkeys is cultured in a Petri dish and inoculated with samples of the vaccine. After incubation, the number of plaques is measured and used as an index of potency. The plaque potency test involves no animals and takes between five and six days compared to the mouse lethality test which takes 21 days.

● The Limulus amoebocyte lysate (LAL) test is an *in vitro* procedure in which subcellular components reveal whether or not therapeutic fluids will induce fever when administered intravenously to human patients. The fever-producing substance, or pyrogen, is a toxic segment of the surface of contaminating bacteria. The active ingredient in this test is obtained from the blood of horseshoe crabs, which are caught in oceans, handled and then released.

The LAL test was recently approved for use by the US government and has already been performed more than a million times. It is beginning to replace the old pyrogen test, which involves a minimum of three rabbits per test, each of which receives experimental fluids that can be so damaging that the animals have to be killed after the procedure. The LAL test is over 100 times more sensitive than the *in vivo* method and is also more economical, convenient and reliable (Wagner and Cooper 1983).

● Tissue culture techniques are at the forefront of basic research in biomedicine, particularly in studies of the immune system. According to the US National Academy of Sciences (NAS 1985), 'major recent advances in our knowledge of the immune system made possible by cell cultures would have been virtually impossible to achieve in intact vertebrates'. The Academy also stated:

> It is clear that the study of *in vitro* antibody responses has led to a major portion of our understanding of immune system responses. Using an *in vitro* system, one can make 300 to 400 cultures from a single mouse. If these same studies were to be conducted *in vivo*, they would require 200 to 400 mice to achieve the same number of observations.

Mathematical modelling

State-of-the-art approaches to biomedical research and testing are increasingly incorporating mathematics into their descriptions of living systems. These mathematical descriptions, or models, are based on existing information about the system under study. Usually they are simplified versions of reality, but are nonetheless helpful in understanding complicated systems, especially those in which several variables influence an outcome.

As an illustration, consider the outcome to be the degree to which various representatives of a certain class of chemicals are toxic. Toxicity is likely to be influenced by several factors, including the size and shape of the chemicals' molecules, the presence of certain reactive groups and the way reactive fragments are linked together. Each of these factors can be represented mathematically by one or more parameters. Toxicity would thus be modelled on the basis of the chemicals' structure, composition and physical–chemical properties. Toxicity data on already-tested compounds could be used to help predict the toxicity of unknown compounds. Models such as these are known as quantitative structure–activity relationships (QSARs) because chemical structure is used to predict activity, in this case, toxicity.

Once formulated, mathematical models must be verified to see if they accurately reflect the relationship under study. In toxicity testing, this verification is known as validation. In the area of research, verification usually involves a procedure known as simulation. Simulations comprise changing one or more parameters in models to determine if the responses are similar to those seen in the living systems. If dissimilar responses are obtained, the model can be refined or entirely reformulated. Simulations usually are conducted with the aid of computers, for tractability. Computer simulations are useful not only in validating models but also in suggesting new mechanisms and hypotheses for further study.

Mathematical modelling is now an integral part of research in many laboratories, particularly in the pharmaceutical industry (Tute 1983). Unfortunately, its more widespread application is hampered by a general lack of mathematical and computer skills among researchers, and the cost of computer equipment and commercially available programs. To help overcome these problems, the National Institutes of Health have recently financed the creation of the Biomedical Simulation Resource at Duke University Medical Center (Anon 1985).

Mathematical models can be replacements for animal experimentation by serving as substitute tests themselves, or by revealing that certain avenues of investigation are not likely to be promising and therefore are not worth pursuing. The extent to which models can substitute for animal experiments depends on how well the models perform during validation. The better the performance, the less the need for back-up animal tests. In toxicity testing, models are likely to bring significant reductions in animal use because existing information from animal studies on thousands of compounds can be applied towards predicting toxicity of closely related compounds that have not been tested. The outlook is not quite as bright when models are applied in new areas of research, given that the results from the simplified models usually need to be checked in the far more complex living system.

Examples of mathematical models follow:

● Mathematical modelling of malaria research illustrates the potential value of modelling in guiding research efforts (Tute 1983). This modelling was a retrospective analysis of results from a large-scale programme that tested potential antimalarial drugs on mice. Development of a structure–activity relationship for a certain class of chemicals synthesised early in the programme showed retrospectively that further research on this class was futile, yet many other chemicals in this class were synthesised and then tested in mice. This analysis suggests that prospective use of mathematical modelling will prevent much futile animal experimentation.

● The potential value of mathematical modelling to cancer research has been illustrated by Dr C. DeLisi and co-workers at the National Cancer Institute (Angier 1983). Their model analysed the response of the immune system to cancer. It revealed that the immune system could not only fight cancer growth but stimulate it as well. Other researchers independently concluded this, but in Dr DeLisi's words, 'if our model had been around ten years ago, it could have predicted what it's taken scientists countless man-hours and animals to figure out. This is the value of mathematical modelling—it comes up with things you might otherwise miss' (DeLisi 1983).

● A computer program developed by 30 scientists at the Los Alamos National Laboratory is an ambitious attempt to duplicate the complex physiological systems of the human body (Duerlinger 1985). The program is known as HUMTRN, short for 'human transport'. It yields information on what happens when any chemically identifiable substance is taken into the human body. HUMTRN is dynamic to the point of being programmed to eat, breathe, perspire, defaecate, grow, develop sexually, age, work and die. A scientist associated with the HUMTRN project has called this program 'the cutting edge of modelling technology'. In one study HUMTRN suggested that, in most kinds of nuclear accidents, teenagers and young adults would be the highest risk group in suffering long-term effects. The developers of HUMTRN refer to this model as the 'research rat of the future'.

● A mathematical model of the LD50 test (Enslein *et al.* 1983, Lander *et al.* 1984) has received a fair amount of attention but has an uncertain future as a potential replacement. The LD50 test is an *in vivo* assay that provides a crude measure of a substance's systemic toxicity. The model of this test predicts oral LD50 values for rats, based solely on a chemical's structure and properties. The predictions, in turn, are based on an analysis of nearly 2000 chemicals that have already been tested in rats. This approach has been criticised as violating an assumption of QSAR analysis by examining chemicals that do not form a congeneric series (see Tute 1983). The merit of this model and similar efforts remains to be established.

Physical–chemical methods

Physical–chemical techniques exploit instruments and chemical procedures, not animals, to analyse the physical and chemical properties of drugs, toxins, body chemicals and other substances. For instance, high-performance liquid chromatographs and mass spectrophotometers are physico-chemical instruments that accurately isolate, identify and measure the amounts of given substances in complex biological mixtures.

● Physico-chemical techniques have replaced animals in assays for vitamins A, B1, B2, B12, C, E, K and nicotinic acid (Smyth 1978). In the case of vitamin D, the new technique involves high-performance liquid chromatography and provides a simpler, quicker and cheaper alternative to the animal bioassay. The bioassay procedure involved inducing rickets in rats and administering vitamin D-rich substances over several weeks—a laborious and time-consuming method (Sharpe 1982).

● High-performance liquid chromatography appears to be on the verge of replacing mice in testing the potency of insulin. In Britain, testing each batch of insulin currently requires 130 mice, down from a high of 600. However, the new non-animal method 'should be acceptable as a satisfactory and reliable replacement for the biological assay of insulin' (Trethewey 1987). If approved by the British Pharmacopoeia, this method could save 33 000 mice per year in Britain alone (Dr Hadwen Trust 1987).

● Physico-chemical techniques have replaced rabbits in human pregnancy testing. Nowadays, one can obtain pregnancy diagnostic kits from the corner chemist. These kits contain simple materials to screen a potential mother's urine or blood for a hormone associated with pregnancy.

Less-sentient organisms

The millions of vertebrate animals that serve as experimental subjects each year have well-developed nervous systems and therefore are more likely to experience pain and suffering from a given procedure than are other types of organisms. Consequently, substituting non-vertebrates for vertebrates, where scientifically valid, would be desirable.

The principle of substituting less-sentient organisms forces us to consider which types of substitutions we would classify as replacements and which as refinements. Refinements would encompass those substitutions involving animals with a significant capacity for pain or suffering, but whose capacity was less than that of the original species. Replacements would encompass substitutions of animals with limited or no capacity for pain or suffering.

This distinction will not always be easy to make. It makes some sense to draw the line at the vertebrate/invertebrate boundary, with substitutions of invertebrates for vertebrates being replacements and substitutions of vertebrates for other vertebrates being potential refinements. This dichotomy is admittedly arbitrary. Some would argue that substituting cold-blooded vertebrates for warm-blooded vertebrates should be considered replacement. Others would argue that some invertebrates have fairly well-developed nervous systems and therefore have more than limited sentience. We should keep these considerations in mind as we adopt the invertebrate/vertebrate dichotomy here.

What about the special case of vertebrate embryos? Under the new Animals (Scientific Procedures) Act 1986 in Britain, vertebrate embryos become 'protected animals' half-way through gestation or incubation. This cut-off point appears to be early enough to ensure that the developing embryo has little or no capacity for pain or suffering. Under the 'half-way' criterion, the CAM test (Leighton *et al*. 1985)— also known as the hen's egg test—cannot be regarded as a replacement alternative. This promising technique is being developed as a possible substitute for the politically charged Draize eye irritancy test, which involves rabbits. Substances are placed on the chorio-allantoic membrane (CAM) of the egg, which has no demonstrable pain fibres, to test for irritancy. Although the method involves embryos beyond the half-way point of development, and therefore perhaps should not be regarded as a replacement alternative, this test would nonetheless be welcomed as a refinement.

Our list of 'alternative organisms' would then include invertebrates and early-stage vertebrate embryos. To this list we can add microorganisms, whether they are technically classified in the animal or plant kingdoms. It might seem foolish to suggest that we even consider macroscopic plants as potential replacements. However, even they have been used to study basic biological processes, such as Mendelian inheritance. Indeed, two Nobel Prizes in Medicine and Physiology have been awarded for research on plants (Stephens 1987). Alternative organisms are being used to develop scores of alternative procedures (Goss and Sabourin 1985, Sabourin *et al*. 1985). Several promising tests have already been developed, many in the field of toxicity testing.

- A simple test for detecting teratogenic chemicals employs the tiny coelenterate hydra. The procedure is based on the observation that chemicals that cause birth defects in animals also tend to disrupt normal development in hydras. This test is currently the most promising alternative screen for teratogens which are usually tested on mammals (Brown 1983).
- Yeast may replace animals in tests to detect substances that cause skin damage in the presence of light (Young 1983). Phototoxins exert their effects after being ingested or applied to the skin. Laboratory animals, particularly hairless mice, are currently used routinely in phototoxicity tests. The alternative test is based on the observation that phototoxins inhibit the growth of yeast in the presence of light. This method yields results that are similar to those from the mouse test for substances that are phototoxic when applied directly to the skin. Further research is needed to corroborate and extend the encouraging results found to date.

● The Ames *Salmonella*/microsome test (Maron and Ames 1983), which employs *Salmonella* bacteria, has been considered to be a classic example of a replacement alternative (e.g. Stephens 1986). It was designed to detect chemical mutagenesis, but because of the association between mutagenesis and carcinogenesis, has come to be applied as a screen for chemical carcinogenesis.

The Ames test showed encouraging promise in predicting carcinogenesis in early evaluations, especially as part of a battery of short-term assays (Campbell and Copeland 1983). However, a recent evaluation (Tennant *et al.* 1987) cast considerable doubt on the suitability of the Ames test, either alone or in combination with other non-animal assays, as a substitute for the traditional rodent carcinogenicity assay. Nonetheless, the Ames test remains a valuable screen for genotoxic chemicals.

Human studies

Human subjects are sometimes overlooked as potential replacements for animals given the ethical constraints of studying human beings directly, without prior animal testing. Indeed, many human studies cannot be considered replacements for animal studies because they are follow-ups of prior research on animals. Such follow-up testing is necessary because results from animal studies may have little relevance to the human condition. The uncertainties of the animal-to-human extrapolation impel scientists to explore fully the potential of human-based research to replace animal experimentation.

Clinical studies immediately come to mind when one thinks of human research, but such research also includes epidemiological and post-mortem studies. Clinical investigation involves the direct study of human volunteers. The focus is on patients who are sick or injured, but healthy volunteers are also studied. For example, volunteers are used extensively to test for possible skin irritation from cosmetics (Smyth 1978). Epidemiological investigation involves the indirect study of human beings; it typically entails statistical analysis of information on large numbers of people to uncover potential relationships between the incidence of some disease or injury and habits such as smoking, drinking and working in certain occupations. These studies are helpful in identifying probable causes of health problems. Similar population analyses are helpful in identifying promoters of good health. Post-mortem investigation involves the study of cadavers donated to science. This research is particularly enlightening in anatomical and transplant studies.

The ethical constraints on human research apply most forcefully to clinical studies of patients. Fortunately, the impact of these fair and proper constraints is slowly but surely diminishing. This is not due to moral decay, but rather to sophisticated new techniques that broaden the horizons of humane clinical study. For example, remarkable new imaging techniques, which can generate visual images of the body's interior with minimally invasive procedures, are now being used for harmless study of the human brain in action. One such technique is positron emission tomography: tiny amounts of radioactive chemicals mark areas of interest in the brain, and a brain scanner detects these chemicals and generates pictures or 'scans' that show the living brain in action. This technique has revolutionised our understanding of Parkinson's disease (Bower 1985, Lewin 1985).

Radical proposals have been advanced by a few physicians and academicians to broaden the horizons of post-mortem studies. The proposals, which raise a host of moral and aesthetic issues, advocate the study of cadavers that are brain-dead but that have been connected to artificial support systems shortly after death (Gaylin 1974, Shane and Daly 1986). Cadavers that had been on such supports before death are also suitable. It is critical to bear in mind that although some of the cadaver's physiological processes have been resumed through artificial intervention, the person—now classified as a 'neomort'—is nonetheless dead.

Such post-mortem studies are said to have the potential to revolutionise research, toxicity testing and education. And from an animal-welfare perspective, they could greatly reduce our reliance on laboratory animals. Support systems are already being employed by the medical community to keep cadavers functioning for less controversial medical or scientific purposes. Practical problems currently make this technology too expensive and complicated for widespread application, although these may be solved in the near future. Clearly, however, a variety of ethical, legal and cultural issues, as well as technical matters, will have to be addressed before brain-dead human beings ever become common research subjects.

To date, perhaps the only study of neomorts was conducted by physicians who implanted the Jarvik-7 artificial heart in brain-dead humans (Kolff *et al.* 1984). They experimented with three surgical implant techniques. They wrote:

> . . . we were confronted with the question of whether or not an artificial heart successfully tested in calves would fit and function in man. But how to proceed in man with some assurance of success? . . . Today it is possible to test the functional capabilities of intrathoracic blood pumps in brain-dead but hemodynamically stable human subjects at no risk, so that it is not necessary to learn the fundamentals of fit and function in patients . . . The relatives of the

deceased subjects have been extremely supportive of our experiments. Their hope is that, through these studies, others may live longer and more comfortably.

Although this study was a follow-up of animal experiments, the clear implication is that, if post-mortem research is sanctioned by society, it could incidentally but significantly reduce our reliance on laboratory animals.

Examples of more conventional human studies are provided below:

● Epidemiological studies, not animal tests, have identified most of the substances known to cause cancer in humans (FRAME 1985). These hazards were identified primarily through occupational association.

● Clinical studies conducted at hospital poison centres are improving treatments for drug-overdose patients (Bennett 1983). These centres are designed so that patients can be studied while being given emergency treatment. One such unit was established at Guy's Hospital in London, where researchers concluded (Goulding and Volans 1982):

> Whilst the data from the animal studies required by regulatory bodies provide some basic information of the mechanism of toxicity and relative toxicity, it cannot be assumed that this information will be entirely relevant for man. Furthermore, whilst these studies may give indications as to the appropriate treatment for acute overdosage, they are unlikely to indicate the efficacy of treatment. Experience gained from a careful assessment of patients suffering from acute overdosage of drugs is potentially much more useful than that obtained from animal tests.

● Clinical and post-mortem studies of patients with Alzheimer's disease have provided a greater understanding of this form of senile dementia (Katzman 1986). Anatomical and biochemical studies have been conducted on brains of deceased patients, as well as on brain biopsies and cerebrospinal fluid from living patients.

● Human beings are becoming a useful alternative in the production of antitoxins against tetanus and other diseases (Smyth 1978). Antitoxins have traditionally been produced in animals, particularly horses. Now human blood can be screened for the presence of useful antibodies, which can be harmlessly harvested. Because human blood is involved, this method requires less safety testing than blood from non-human species.

Other methods

Other techniques or systems could serve as replacements for animals in research. These include computer-aided drug design, veterinary patients and mechanical models.

Computer-aided drug design Discovering new drugs is largely a trial-and-error process, costly in terms of time, money and animals. According to one estimate, 7000 to 8000 novel compounds are screened for every one that reaches the medical marketplace, and this process takes eight years, on average (Wright 1983). Fortunately, methods are being developed to replace this shotgun approach with the more directed approach of computer-aided drug design. Three-dimensional computer graphics and the theoretical field of quantum pharmacology are being employed to design drugs with particular specifications or to improve the specificity of existing drugs. These efforts are based on the lock-and-key mechanism of drug action; that is, drugs must be the right shape and composition in order to 'dock' with their targets and trigger their effects. Colour graphics help visualise this process.

The young field of drug design is maturing. Many 'drug designers' now work in academic and commercial laboratories. Several have been included on new drug patents for aid in discovering drugs (Anon 1983). Their methods have been applied in research on cancer, cardiovascular disease, sickle-cell anaemia and other areas (Wright 1983, Anon 1984, Freiherr 1987). Efforts such as these hold great promise for reducing animal use by revolutionising the process of drug discovery.

Veterinary patients Just as clinical studies of humans could replace some uses of laboratory animals, so too could clinical studies of animals. Animals are susceptible to many of the illnesses and injuries that plague humans. Animals that are already sick could be studied while undergoing treatment, and the resulting knowledge could benefit human health. Of course, the primary concern in these cases should be the animals. Nonetheless, clinical studies of animals could reduce the number of laboratory animals that are deliberately sickened or injured in experiments.

Glickman and Domanski (1986) have drawn attention to epidemiological studies of cats and dogs as possible alternatives in human health-risk assessment. Schwabe argues that both clinical and epidemiological studies of animals are being virtually overlooked as potential resources for understanding human diseases. The relevance of spontaneously occurring diseases in animals to medical research on humans is unappreciated. A consequence of this, according to Schwabe, is that most of the research in comparative medicine conducted by physicians is focused upon the potentially least rewarding approach to animal diseases, namely, artificially induced rather than spontaneous diseases.

Michael W. Fox, in recounting Schwabe's view, advocates greater collaboration between veterinary and medical researchers (Fox 1986).
Mechanical models Animals are sometimes used to study the effects of accidents (such as vehicle crashes) and specific injuries (such as burns). Mechanical models are being developed that might replace animals in these studies. For example, an artificial neck developed by General Motors has been used in car-crash simulation tests (Rowan 1979), and a human simulator known as Thermoman has been employed in testing potential burn risks with different garments (Pratt 1980).
Miscellaneous methods Another potential alternative to some experimental studies on animals is the ethological study of animals in the wild (Pratt 1980).

Not all applications of various replacement methods discussed above actually serve as true alternatives to animal experiments. In some studies, these methods are employed because *in vivo* animal experiments are incapable of answering the question at hand. Many molecular biology studies, conducted *in vitro*, are examples. Such studies are testaments to the power of replacement techniques and should be welcomed for contributing to biomedical knowledge in ways that traditional animal methods cannot. Nonetheless, they are best not classified as alternatives.

HISTORICAL APPLICATIONS

Most of the methods we now label as 'replacements' existed prior to Russell and Burch's (1959) original formulation of the alternatives approach (although they have been developed for application to many more areas of research since then). Scientists employing these methods before 1959 probably were motivated almost exclusively by scientific— not humane—concerns, and this is probably true even for today's alternatives-based research, although humane concerns are starting to play a more influential role.

The historical significance of replacement techniques can be gauged from the contribution of these methods to research that has been highly regarded by the scientific community. The best barometer of such regard—though not an infallible one—is the awarding of the Nobel Prizes in Medicine and Physiology. According to the prestigious US National Academy of Sciences (NAS 1985), these prizes are generally believed to recognise research 'of the highest caliber, the most enduring influence, and the most importance to biomedical science'.

The Humane Society of the United States estimated that replacement methods made major contributions to the prize-winning research for fully two-thirds of all awards up until 1985, or a total of 50 awards (Stephens 1987). All of the major types of replacement methods were represented among these projects, with *in vitro* techniques outshining all other approaches.

The 50 studies in the replacement category were further classified into those that could be regarded as alternatives to animal-based projects and those that could not (see preceding section). The projects fell about equally in both subcategories (24 vs. 26, respectively).

The award-winning projects in either subcategory form a diverse collection. Some projects were of practical value in the fight against disease. For example, the successful cultivation of the polio virus in tissue culture (awarded the Nobel Prize in 1954) paved the way for the development of safe and effective vaccines against polio. Other award-winning projects using replacement techniques made significant contributions to basic biology, such as the discovery of the interaction between tumour viruses and the genetic material of cells (1975). Other projects developed techniques such as radioimmunoassay (1977) that have rapidly become invaluable in biomedical research.

The foregoing analysis, which is presented in detail by Stephens (1987), is not meant to imply that animal experimentation played a trivial role in research awarded the Nobel Prize. However, it does serve to temper the claims of the overarching importance of animal research, made by those who have analysed such prizes. These analyses, by the National Society for Medical Research (NSMR undated) and Leader and Stark (1987), have fundamental flaws. First, they did not distinguish between studies that used traditional *in vivo* methods and those that used alternative methods (*in vitro* studies, clinical veterinary studies and ethological studies). Secondly, no assessment was made of the importance of intact vertebrates in research projects that also involved replacement techniques. Thirdly, the NSMR list includes several projects on invertebrates such as fruit flies, although public concern and legislation focus on vertebrates. When these factors are taken into consideration, Nobel Prizes provide far less support for the historical importance of animal experimentation; rather, they provide a surprisingly high level of support for the importance of research employing replacement methods.

MISSED OPPORTUNITIES

The number of major contributions that replacement techniques have made to Nobel Prize-winning research is astonishing, as this approach has yet to be embraced as a guiding principle of biomedical research.

The driving force behind the adoption of alternative methods in this research has been scientific innovation. Yet the emergence of the alternatives approach as an additional force, albeit a weak one to date, is cause for optimism about the pace of future progress in substituting replacement methods for current uses of animals.

As striking as the results of the Nobel Prize analysis are, even more awards would have gone to projects that used replacement techniques if not for the traditional emphasis on *in vivo* vertebrate studies in biomedical research. For example, many researchers were sceptical of tissue culture systems in the early days of this technique's existence. If not for this scepticism, tissue culture 'might have been used to discover many of the vitamins, amino acids, and hormones' according to the US National Academy of Sciences (NAS 1985). Missed opportunities to employ replacement methods probably occurred to some degree in research at all levels of sophistication, not only in Nobel Prize-calibre work. Such missed opportunities translate into unnecessary use of animals.

Stratmann *et al.* (1987, p. 10) advance a different point of view, claiming that it is 'somewhat specious' to argue 'that very many animals have been wasted in the past and alternative methods, such as tissue culture, would have saved them if only scientists had been prepared to use them'. They defend past scientists by noting that the prevailing state of knowledge was limited. This is indeed true, but widespread implementation of any scientific innovation frequently lags behind its initial development by several years or more, even when such innovations are in the interest of the practitioner (Baker 1986). In the case of alternative methods, implementation probably would have been quicker if humanitarian concern for animals in laboratories had been greater.

Missed opportunities to employ replacement methods are reflected in funding patterns. The US National Academy of Sciences assessed the value of research involving various 'model systems' and then examined the corresponding funding levels of research on those systems (NAS 1985). The Academy was particularly interested in research on non-mammalian species. Unfortunately, it did not emphasise the *in vivo/in vitro* distinction. It reported that studies of micro-organisms and invertebrates (that is, less-sentient organisms) as well as studies of 'lower vertebrates' have made great strides in our understanding of biology and medicine. Yet, the report noted, 'the proportion of [National Institutes of Health] resources that supports research in this area may be small in comparison to the resources dedicated to research with mammals.'

The apparent over-funding of research on mammals stems at least partly from what Russell and Burch (1959) termed the 'high-fidelity fallacy'. A model system, in either animals or alternatives, is of high fidelity to the extent that it resembles human beings. Chimpanzees are high-fidelity models; bacteria are low-fidelity models. Models can have low fidelity but nevertheless be more valuable than high-fidelity ones in certain cases. This is because low-fidelity models can be better discriminators of the particular response under study. For example, horseshoe crabs in the LAL test (see above) are replacing rabbits in pyrogen testing. Horseshoe crabs happen to be better than rabbits in identifying substances that induce fever in humans, despite the fact that horseshoe crabs are lower-fidelity models of humans.

A failure to consider a model's discriminating ability can lead to the high-fidelity fallacy. This fallacy ignores discrimination by stating uncritically that, in general, models should have high fidelity. In practice, this fallacy leads to overuse of mammals, given their relatively high fidelity to humans.

Of course, the judicious use of mammals can make sense scientifically (although not always economically). However, their uncritical and unnecessary use at the expense of other model systems impedes progress in the development and implementation of replacement methods.

INCENTIVES TO EMPLOY REPLACEMENT METHODS

Scientific and practical incentives—apart from any ethical considerations—have been paramount in the development and implementation of replacement methods. Each replacement method has scientific advantages that recommend its application in certain circumstances (see section entitled 'Types of replacement' above), and many of these applications tend to be more rapid and economical than traditional animal experiments.

These practical and scientific incentives to seeking alternatives, though quite influential, may seem secondary to the humanitarian incentive. Indeed, the conceptual unity of the alternatives approach is based on humanitarian concern. Nevertheless, the variety of reasons for seeking alternatives is an asset. It fosters co-operation between those whose primary motivations may differ. Many scientists care about animals and may therefore be alert to the possibility of developing or implementing alternative methods in their own laboratories. They may investigate alternatives not only for the sake of the animals, but also as a goodwill gesture to an increasingly outspoken and politically astute public.

SOURCES OF FUNDING

In Britain and the United States, few conventional sources of funding for biomedical research have earmarked money specifically for the development and validation of alternatives to animal-based methods. To be sure, these sources do fund research that involves replacement methods and, indeed, are by far the biggest underwriters of alternatives-based research. The problem is that special funding is needed to expand the capabilities of replacement techniques and thereby remove any existing barriers to their more widespead application.

What is available from conventional sources? While the British research councils still have no funds established for alternatives research, the US National Institutes of Health has a modest programme on alternatives. In 1985, Congress directed the NIH to establish a plan for research into replacements, reductions and refinements. The plan was also to include the development of such methods that have been found to be valid and reliable, and the training of scientists in their use. The plan was announced in early 1987. Unfortunately, no special fund nor grant-review mechanism was created for this research. These features are likely to diminish the plan's impact.

In reponse to this inadequate funding from conventional sources, several animal protection organisations have established funding programmes of their own, some of which, in Britain, date back to the 1960s. Table 7.1 lists several such sources. Other organisations offer monetary awards for the development of viable alternatives (Table 7.1). Although these organisations collectively dispense far less money than do conventional sources, they provide valuable seed money, highlight the humanitarian nature of the search for replacements, and enhance the legitimacy of this search in the eyes of many scientists.

Private industry has established funding programmes in response to calls from animal protectionists to develop products without testing them on animals. Revlon, Avon and numerous other companies have donated hundreds of thousands of dollars to prestigious institutions such as Rockefeller University and Johns Hopkins University to develop alternatives to the Draize eye irritancy test (Spira 1985). Other companies have targeted other animal tests either through similar donations or through in-house research efforts. Industry donations to Johns Hopkins University financed the creation of the Center for Alternatives to Animal Testing, which currently funds a variety of in-house and extramural projects.

SCIENTISTS' ATTITUDES

The attitudes of members of the animal research community towards the alternatives approach range from enthusiastic to hostile. Many

Table 7.1   Some Sources of Financial Support for the Development of Alternatives

(a) Organisations established to finance development of alternatives

Organisation	Reference
American Fund for Alternatives to Animals in Research (USA)	AFAAR (1978)
Dr Hadwen Trust for Humane Research (UK)	Smyth (1978)
FRAME (Fund for the Replacement of Animals in Medical Experiments) (UK)	Smyth (1978)
Fund for Experimental Animal-Free Research (Switzerland)	Anon (1987a)
Humane Research Trust (UK)	Smyth (1978)
International Foundation for Ethical Research (USA)	IFER (1987)
Irish Anti-Vivisection Society Humane Research Fund (Ireland)	—
Lawson Tait Trust (UK)	Smyth (1978)
Lord Dowding Fund for Humane Research (UK)	Smyth (1978)

(b) Organisations offering monetary awards for alternatives development

Organisation	Reference
Canadian Society for the Prevention of Cruelty to Animals	Rowsell and McWilliam (1986)
Doerenkamp and Zbinden Foundation (Switzerland)	Rowan (1987)
European Federation of Pharmaceutical Industries Association	—
Hutzenlaub Research Award	Smyth (1978)
Jorio Rusticelli Award (Italy)	Rowan (1987)
Marchig Award (Switzerland)	Anon (1987b)
Millenium Guild (USA)	OTA (1986)
Wankel Prize (West Germany)	Rowan (1987)

researchers, laboratory animal veterinarians and others in this community are somewhat resistant—if not antagonistic—to the alternatives approach. They hesitate even to speak of 'alternatives' or 'replacements', referring instead to 'adjunctive methods', an epithet which implies that all forms of research, including human studies, are ancillary to animal experimentation.

Part of this antagonism stems from the general human tendency to resist change, and has nothing to do with the controversy surrounding vivisection. Many of today's biomedical scientists were trained using traditional *in vivo* methods and, understandably, are reluctant to switch to unfamiliar techniques. Similarly, the entire animal research industry, including the animal breeders, cage manufacturers, feed suppliers and related companies such as the pharmaceutical manufacturers, are somewhat reluctant to change the *status quo*. The replacement of animal-based methods poses an inconvenience, if not a financial threat, to all components of this industry.

Resistance to change is a powerful force. Nevertheless, the vivisection controversy itself lies at the heart of many scientists' antagonism towards the alternatives approach. Scientists see themselves as battling with animal protectionists for the hearts and minds of the public. They view present alternatives as far from the ultimate goal of complete replacement or they view the goal itself as unattainable. Accordingly, they are hesitant to refer to existing techniques as alternatives or replacements for fear of encouraging overenthusiasm for these methods and losing public support and government backing for their own work.

Personal psychology may play a small but significant role in some scientists' antagonism towards alternatives. A number of scientists probably think that if they agree with their critics that alternatives should be pursued, they would be admitting that something is wrong with vivisection. And for those scientists who do feel that something is wrong with vivisection, their guilt might be exacerbated if they conceded not only that alternative methods should be pursued, but also that in some cases they already exist. One also gets the impression that even if some scientists actually shared the animal protectionists' enthusiasm for the alternatives approach, they would not express it. There is a hesitancy to agree to certain aspects of a critic's platform for fear of giving credence to the rest of that platform.

Of course, antagonism towards the search for alternatives does not necessarily translate into antagonism towards existing alternatives. Stratmann *et al.* (1987, p. 8) noted that they 'have yet to meet scientists who would not prefer to use an alternative rather than a whole animal if it were possible to do so and if they were sure of obtaining the necessary result'.

Whatever the reasons for the substantial resistance to alternatives among scientists, such resistance is decreasing and is likely to fade as hard-line defenders of vivisection retire and are replaced by a more open-minded generation of scientists, and as more scientists join with reformers in good-faith efforts at humanitarian progress.

THE FUTURE

The field of alternatives research, though young, is already beginning to show signs of coming of age. Funding opportunities and publication outlets are expanding, and scientific conferences on alternatives are becoming more frequent and better attended. As the field gains widespread respectability, it will gain further momentum.

That the development and implementation of replacement methods will advance is inevitable. The uncertainty lies in the pace of this progress. Animal protectionists will be continuously exerting pressure on scientists and others to quicken this pace, whereas scientists probably will call for a more conservative effort to minimise effects on ongoing research.

Although scientists themselves will be the ones working on replacements in the laboratories, others have an important role to play given their involvement in the research process. Funding agencies, for example, will probably be the most important determinant of the pace of change. These agencies clearly should provide more support for alternatives research. The US government's own scientific think tank, the National Academy of Sciences, concluded that the government was under-funding certain methods of research (NAS 1985), and these methods are ones that animal protectionists would label as replacements. The problem was, and continues to be, that research on traditional mammalian animal models has been over-funded. The Academy had this advice:

> Proposals for the study of invertebrates, lower vertebrates, micro-organisms, cell- and tissue-culture systems, or mathematical approaches should be regarded as having the same potential relevance to biomedical research as proposals for work on systems that are phylogenetically more closely related to humans. Support should be given to good research without taxonomic or phylogenetic bias on the part of the sponsor . . .

Funding patterns in the United States and elsewhere have not kept pace with the scientific potential of alternative methods nor the public's concern for animals in laboratories. Research employing alternative techniques *is* being funded, but in an unco-ordinated manner. Moreover, little effort is being made to expand the capabilities of existing alternative methods. A great need exists for well-funded, co-ordinated programmes for research and development of alternatives.

In addition to funding agencies, another influential party is the regulatory community. Many regulatory agencies have requirements for toxicity testing on animals. In some cases the requirements are formal, in others they stem from accepted practice. The task force of the European Chemical Industry Ecology and Toxicology Centre (ECIETC 1985) recently evaluated acute toxicity testing; one of its recommendations was that regulatory agencies encourage the use of alternatives 'to prevent unnecessary use of animals'. Requirements for outmoded and scientifically questionable tests such as the LD50 need to be changed; scientists should call on regulatory agencies to encourage the development and use of alternatives-based tests. International efforts are needed to preclude animal testing to satisfy requirements in foreign markets when such testing is not required or condoned in the home country.

The public is another major player in hastening the development and implementation of alternatives. This can be accomplished by campaigning *for* alternatives or, indirectly, by campaigning *against* animal testing or research. In some European countries, efforts are under way to promote steady reductions in animal use. The success of these efforts is bound to stimulate progress on alternatives, given the truth in the aphorism that necessity is the mother of invention.

The late Sir Peter Medawar, a Nobel Laureate in medicine and physiology, predicted a decline in animal use in the absence of public pressure. In defending animal research, he noted (Medawar 1972):

> . . . this does not imply that we are for evermore, and in increasing numbers, to enlist animals in the scientific service of man. I think that the use of experimental animals on the present scale is a temporary episode in biological and medical history . . .

Despite the significant roles of funding agencies, regulatory agencies and the public, scientists are the ones who have the technical capability of hastening the day when the exploitation of animals is no longer regarded as necessary for research. Through their research on replacement methods and through their interactions with sceptical colleagues, they will usher in a new era of research techniques.

An expanded role for alternative techniques can transform biomedical research from an animal-centred enterprise to a human-centred one. Human-centred research would emphasise ethical and sophisticated clinical, epidemiological and post-mortem studies, as well as *in vitro* and physico-chemical studies of human material and mathematical modelling of human-derived data. Eventually animals may no longer be exploited even in the limited role of providing cells, tissues or organs. Today's alternative techniques may become tomorrow's mainstream techniques.

ACKNOWLEDGEMENTS

I thank Drs M. Balls, D. Fanfarillo and A. Rowan for commenting on earlier drafts of this manuscript.

REFERENCES

AFAAR (American Fund for Alternatives to Animal Research) (1978). Newsletter, *AFAAR News*, Summer

Angier, N. (1983). The electronic guinea pig, *Discover*, September, 77–80

Anon (1983). One of the first three, *Bull. Lord Dowding Fund*, Spring, 12

Anon (1984). Computer models aid drug designers, *New Sci.* , August, 23

Anon (1985). Mathematical modelling help, *Lab. Anim.* , **14** (September), 12

Anon (1987a). Fonds fur versuchstierfreie Forschung (FFVFF), *ATLA*, **15**, 140–1

Anon (1987b). First Marchig award goes to FRAME, *FRAME News*, Summer, 1–2

Baker, F. (1986). Diffusion in new methodology. Paper presented at symposium *In Vitro Toxicology: Approaches to Validation*, Johns Hopkins Center for Alternatives to Animal Testing, Baltimore, MD, 14–15 April

Bennett, P. N. (1983). In Balls, M., Riddell, R. J. and Worden, A. N. (eds), *Animals and Alternatives in Toxicity Testing*, Academic, London, pp. 435–9

Blue Sheet (1985). 20 February. FDC Reports, Chevy Chase, MD

Bower, B. (1985). Tracking the roots of Parkinson's disease, *Sci. News*, **128**, 212

Brown, N. A. (1983). In Balls, M., Riddell, R. J. and Worden, A. N. (eds), *Animals and Alternatives in Toxicity Testing*, Academic, London, pp. 214–8

Campbell, D. J. and Copeland, K. F. T. (1983). *Report on Alternatives to the Use of Animals in Research*, Animal Welfare Foundation of Canada

DeLisi, C. (1983). Quoted in Angier (1983)

Duerlinger, J. (1985). Computerized human body advances ecology research, *New York Times*, 12 November

Enslein, K., Lander, T. R., Tomb, M. E. and Craig, P. N. (1983). *A Predictive Model for Estimating Rat Oral LD50 Values*, Benchmark Papers in Toxicology, Vol. 1, Princeton Scientific, Princeton, NJ

ECETOC (European Chemical Industry Ecology and Toxicology Centre) (1985). *Acute Toxicity Tests, LD50 (LC50) Determinations and Alternatives*, Monograph, No. 6 (May)

Fox, M. W. (1986). *Laboratory Animal Husbandry: Ethology, Welfare and Experimental Variables*, State University of New York, Albany, NY

FRAME (Fund for the Replacement of Animals in Medical Experiments) (1985). Alternatives to animal experiments, *Anim. Int.* Summer/Autumn, 6–7

Freiherr, G. (1987). Genes, drugs and computers: the high-tech revolution, *Res. Resources Reporter*, **11** (September), 1–5

Gaylin, W. (1974). Harvesting the dead, *Harper's*, **52** (September), 23–30

Glickman, L. T. and Domanski, L. M. (1986). An alternative to laboratory animal experimentation for human health risk assessment. Epidemiological studies of pet animals, *ATLA*, **13**, 267–85

Goss, L. B. and Sabourin, T. D. (1985). Utilization of alternative species for toxicity testing: An overview, *J. Appl. Toxicol.*, **5**, 193–219

Goulding, R. and Volans, G. N. (1982). Quoted in Sharpe, R. (1982). Poisons information. *Bull. Lord Dowding Fund*, Autumn, 10

Griffiths, B. (1984). Animal cell cultures in medicine, *Bull. Lord Dowding Fund*, Spring, 5–12

Dr Hadwen Trust for Humane Research (1987). *Replacing Animal Experiments* (pamphlet), Dr Hadwen Trust for Humane Research, Herts

IFER (International Fund for Ethical Research) (1987). *IFER Newsl.* , **1** (no. 1)

Katzman, R. (1986). Medical progress, Alzheimer's disease, *New Engl. J. Med.*, **314**, 964–73

Kolff, J., Deeb, M., Cavarocchi, N. C., Riebman, J. B., Olsen, D. B. and Robbins, P. S. (1984). The artificial heart in human subjects, *J. Thor. Cardio. Surg.*, **87**, 825–31

Lander, T., Enslein, K., Craig, P. and Tomb, M. (1984). In Goldberg, A. M. (ed.), *Alternative Methods in Toxicology*, vol. 2, *Acute Toxicity Testing: Alternative Approaches*, Mary Ann Liebert, New York, p. 183

Leader, R. W. and Stark, D. (1987). The importance of animals in biomedical research, *Persp. Biol. Med.*, **30**, 470–85

Leighton, J., Nassaurer, J. and Tchao, R. (1985). The chick embryo in toxicology: an alternative to the rabbit eye, *Food Chem. Toxicol.*, **23**, 293–8

Lewin, R. (1985). Clinical trials for Parkinson's disease?, *Science*, **230**, 527–8

Maron, D. M. and Ames, B. N. (1983). Revised methods for the Salmonella mutagenicity test, *Mut. Res.*, **113**, 173–215

Medawar, P. B. (1972). Quoted in Rowan (1981), p. 528

NAS (National Academy of Sciences) (1985). *Models for Biomedical Research*, National Academy Press, Washington, DC

NSMR (National Society for Medical Research) (undated). *What Do These Nobel Prize Winners Have in Common?* (pamphlet), NSMR, Washington, DC

OTA (Office of Technology Assessment) (1986). *Alternatives to Animal Use in Research, Testing, and Education*, OTA, Washington, DC

Pratt, D. (1980). *Alternatives to Pain in Experiments on Animals*, Argus Archives, New York

Rowan, A. N. (1979). *Alternatives to Laboratory Animals*, Institute for the Study of Animal Problems, Washington, DC

Rowan, A. N. (1981). In Sperlinger, D. (ed.), *Animals in Research: New Perspectives in Animal Experimentation*, Wiley, Chichester, pp. 257–83

Rowan, A. N. (1987). Alternatives ready for another resurgence, *Our Anim. Wards*, October, 5

Rowsell, H. C. and McWilliam, A. A. (1986). The search for alternatives: the Canadian initiative, *ATLA*, **13**, 210

Russell, W. M. S. and Burch, R. L. (1959). *The Principles of Humane Experimental Technique*, Methuen, London

Sabourin, T. D., Faulk, R. T. and Goss, L. B. (1985). The efficacy of three non-mammalian test systems in the identification of chemical teratogens, *J. Appl. Toxicol.*, **5**, 227–33

Shane, H. G. and Daly, W. J. (1986). How the dead can help the living, *Futurist*, January–February, 24–6

Sharpe, R. (1982). Continuing progress, *Bull. Lord Dowding Fund*, Autumn, 28

Smyth, D. H. (1978). *Alternatives to Animal Experiments*, Scolar Press and Research Defence Society, London

Spira, H. (1985). In Singer, P. (ed.), *In Defence of Animals*, Basil Blackwell, Oxford

Stephens, M. L. (1986). *Alternatives to Current Uses of Animals in Research, Safety Testing, and Education*, The Humane Society of the US, Washington, DC

Stephens, M. L. (1987). In Fox, M. W. and Mickley, L. D. (eds), *Advances in Animal Welfare Science 1986/1987*, Martinus Nijhoff, The Hague

Stratmann, G. C., Stratmann, C. J. and Paxton, C. (1987). *Animal Experiments and Their Alternatives*, Merlin Books, Braunton, Devon

Tennant, R. W., Margolin, B. H., Shelby, M. D., Zeiger, E., Haseman, J. K. , Spalding, J., Caspary, W., Resnick, M., Stasiewicz, S., Anderson, B. and Minor, R. (1987). Predicting carcinogenicity in rodents from *in vitro* genetic toxicity assays, *Science*, **236**, 933–41

Trethewey, J. (1987). Moving away from animal tests for the standardization of insulin formulations, *TIPS*, **8**, 287–8

Tute, M. S. (1983). In Balls, M., Riddell, R. J. and Worden, A. N. (eds), *Animals and Alternatives in Toxicity Testing*, Academic, London, pp. 337–66

Wagner, Jr, H. and Cooper, J. (1983). Horseshoe crab provides alternative to rabbit bioassay, *Newsl. Johns Hopkins Center Altern. Anim. Test.*, Baltimore, MD, Fall, 1–2

Wright, P. (1983). *The Times*, London; reprinted in *Bull. Lord Dowding Fund*, Autumn, 20

Young, A. R. (1983). In Balls, M., Riddell, R. J. and Worden, A. N. (eds), *Animals and Alternatives in Toxicity Testing*, Academic, London, pp. 327–9

8. The Scientist's Responsibility for Refinement: A Guide to Better Animal Welfare and Better Science

David B. Morton

'Those who think that science is ethically neutral confuse the findings of science, which are, with the activity of science, which is not.'

Bronowski, J. (1956). *Science and Human Values*, Harper & Row, New York

INTRODUCTION

The aim of this essay is to try to highlight ways that the use of animals in experimental work can be modified to reduce animal suffering and the number of animals used. In particular, new licensees are addressed but perhaps some of the following points may also be of use to others. (This essay, so far as the legal aspects are concerned, is specifically aimed at those who work within the UK.) Some of these topics will be covered in initial discussions with supervisors, licence sponsors, or the person in charge of the animal facilities in which the work will be carried out. The details will obviously depend on the type of research work and the experience of the individual, but there will inevitably be an underlying recurrent theme: that of not causing animal suffering which is not strictly necessary as part of the scientific objective. This essay describes how research work can be refined so that better animal welfare, and often better science, can result. It is also perhaps interesting to reflect that, if research on animals was never painful and did not involve confining them, it is doubtful whether the public would be as concerned, or respond in the same way to other people's (anti-vivisectionists') concerns. Andrew Rowan's book makes interesting reading for those who wish to delve deeper into the use of animals in science (Rowan 1984).

Many years ago William Lane-Petter posed the following questions [*my paraphrasing*], which are still as relevant today:

● Is the problem worth solving?
● Is an animal the best model?
● Must the animal be conscious for the experiment?
● If it must be conscious, can the adverse effects be reduced?
● Is the number of animals appropriate?

SOME CONSIDERATIONS WHEN USING ANIMALS FOR SCIENTIFIC PURPOSES

Researchers must be aware of their many responsibilities when working with animals if for no other reason than, so far as we can tell, animals show very similar signs to ourselves when in pain, distress and discomfort; that is they too have the ability to 'suffer', as pointed out by Jeremy Bentham (1789). It is difficult to prove that animals suffer, just as it is difficult for *us* to prove to each other that one of us has a headache or feels some discomfort. But at least we can sympathise with what the other is describing as at some time we too may have suffered something similar in the past. With animals, however, we have to go one step further and bridge a species gap.

Bridging this gap must become easier as we learn more about the natural behaviour and physiology of a particular species, or breed/strain of animal, or indeed about that individual animal. It helps us empathise with and understand any suffering that may be occurring. Some argue that even keeping animals in cages is a form of suffering and, while it is certainly a restriction of movement, I am not sure if that *per se* is suffering, though taken to extremes it can be so (e.g. chimpanzees in small cages). Some species have been bred in captivity for decades and consequently there will have been a selection for those animals that can adapt more easily to the laboratory environment (similar to the domestication of companion and farm animals). Practically, however, all such animals have their natural pattern of behaviour restricted. There is some evidence to show they never lose their natural instincts, even though they have overtly adapted to the new surroundings, and we should be looking more carefully into their unfulfilled ethological requirements (i.e. their behavioural *needs* as opposed to their behavioural *repertoire*—see Dawkins (1980)). Perhaps from a humanitarian point of view we should try to make sure that animals can carry out some non-essential behaviours, as they can undoubtedly experience a state of 'happiness'; take for example a dog enjoying a game with a ball. It is certainly not a behavioural need, but nevertheless toys and the like can relieve the monotony, as we are realising in zoo and farm animals. Such 'extras' can prevent, modify and stop the self-mutilations and stereotypic behaviour patterns (sometimes called 'vices') that are seen.

Limitations on movement and behaviour patterns may cause stress in animals, which if prolonged or excessive may induce a state of *distress*, i.e. a stress that adversely affects an animal's normal physiological functions and reactions. There is then also scientific merit in avoiding such situations, as stressed animals will not make reliable experimental models (unless the study deliberately involves the state of stress). There is much anecdotal and scientific evidence that stress and distress can confound scientific results, e.g. by predisposing animals to disease, and accelerating tumour growth and disease conditions (Gartner *et al.* 1980, Moberg 1985, Weber and van der Walt 1973). Interestingly, it is possible to carry out the reverse: animals handled carefully or deliberately kept under less stressful husbandry conditions resist or slow the experimental induction of disease states (Nerem *et al.* 1980, Riley 1981, Kaplan *et al.* 1983).

I do not wish to discuss the issues surrounding whether animals have rights, but I do not believe that many, if any, scientists would use an animal if there was a non-sentient alternative—animals are used because they provide invaluable information about the whole animal, something that *in vitro* techniques cannot do (but see essay by Stephens in this volume). However, we have a duty to animals, as well as to other humans, not to cause them needless suffering (given that one has personally decided it is ethical to use animals for the proposed scientific purpose, and that it is legal so to do). So how does one go about reducing the adverse effects that may be caused by the research work one wishes to carry out? How can it be ensured that only the minimum amount of suffering is inflicted when conducting experiments on animals? One has always to appreciate that animals, like sophisticated electronic equipment, are sensitive complex systems, but unlike this machinery they are sentient as well as living. Animals, too, can give misleading results unless they are handled carefully and unless it is understood what causes variation in them. It is for this reason that we try to standardise their environment and their genetic background as much as possible.

Search the literature and consult colleagues

A literature search should be carried out in the local departmental or institutional library on the models that are available to answer the scientific question posed. This can be done by means of computerised literature searches and then going on to journals which abstract scientific articles (e.g. *Biological Abstracts, Chemical Abstracts*). Finally, maintaining awareness of developments in the field and gaining new ideas can all be fostered by reading relevant journals, scanning *Current Contents* and the many 'newspapers', going to meetings and discussing the research with colleagues.

When reading the literature one must remember that, regrettably, some scientific papers do not always mention the techniques or dosages which have been used and failed. Thus a final group size of 10 may represent the only survivors from a much larger original group. When results are written up, it is unethical not to mention the failures as well as the successes, in order to prevent other animals suffering needlessly in another laboratory where they are trying to use the same experimental model. The current trend of the anti-vivisection groups and other individuals (e.g. Silcock 1986) to scan and select papers from the scientific literature for criticism does not encourage scientists to report accurately in this manner, which is unfortunate, but perhaps it does not stop one telephoning or corresponding directly if there seems to be extra information to be gained. These unofficial scrutineers sometimes make good points which emphasise the need for a fuller debate on these issues; abolitionists are rarely productive allies.

Look carefully at the alternatives

Alternative methods of achieving the scientific aim must be considered, especially those that do not involve the use of living animals. Sometimes these alternatives can be more sensitive and answer scientific questions in a more direct manner because the variables can be controlled more precisely. For example, one might initially use cell or organ culture systems, though it is quite likely that at some later stage in the research programme the whole animal will have to be used. Narrowing the field of research down to the subcellular or molecular level can be extremely valuable in answering basic questions but the whole body is not like that in reality—it is a complex mixture of molecules of a complex structure, it has intracellular organelles, cells and organs, and, at another level, it consists of individuals influenced by social cultures and interacting with each other. A moment's thought about the study of pain at all these levels will illustrate the point.

The Fund for the Replacement of Animals in Medical Experiments (FRAME) has started a data bank on the use of *in vitro* techniques in toxicology and this may have some up-to-date information on *in vitro* techniques. FRAME also publishes a journal—*ATLA* (*Alternatives to Laboratory Animals*)—whose contents include regular bibliography selections, news, views and reviews, as well as original papers on alternative research techniques and their validation.

I use here the word 'alternative' to mean any one of the following:

- Improving (minimising any adverse effects on the animal) the way in which a living animal is used—known as refinement (see Smyth 1978, Hampson and Silcock 1985).
- Replacing the use of living animals—possibly by killing them and using the tissues, or by using permanent cell lines (the latter can be construed as a true replacement—especially when human cells are used).
- Reducing the number of animals used.

Refinement, replacement and reduction form the three 'R's put forward by Russell and Burch (1959).

Journals such as *Laboratory Animals*, *Laboratory Animal Science* and *ATLA*, newsletters from the Institute of Laboratory Animal Resources, Scientists' Center for Animal Welfare and Canadian Council on Animal Care Resource, as well as the specialist journals within an academic discipline, will contain papers on refinement techniques and are worth scanning regularly. For other examples of refinement see Silcock (1986), who quotes several specific cases where she believes refinement of experimentation could have been carried out.

Having decided there are no *in vitro* alternatives or replacements to the proposed use of animals, and that the model chosen is the best way of achieving the scientific purpose, there are still several points in relation to refinement to consider before starting an experiment.

FACTORS AFFECTING THE CHOICE OF AN ANIMAL MODEL AND ENVIRONMENTAL CONDITIONS

The following sections deal with some of the common choices that are available to research scientists, and provide some basic information on the way animals are kept and how laboratory animal science can enable reduction and refinement of procedures as well as contributing to more reliable research.

Species

Species selection is likely to be based on work that has previously been carried out and the original choice probably took account of the normal physiology and anatomy of that species. It goes without saying that the results obtained will have particular relevance to the species on which the research was carried out, but for many types of study there is an underlying belief that the results will also be applicable to other species, in particular to people (e.g. toxicity testing, research and development of new drugs), and while this may not always be true, far more often than not, it is (but see essay by Sharpe in this volume). The commonly used laboratory animals are rodents (rats, mice and guinea-pigs) and lagomorphs (rabbits), but many other species are used in research. It is always important that the species chosen will be that which will give the best probability of achieving valid scientific results. This often is not predictable, but if it is, it may mean using dogs, cats or primates. However, with the larger animals and with those species which cause particular public concern (dog, cat, primate, horse), the cost in monetary terms is often high and, practically, one may have to use the traditional, smaller laboratory species. Other exceptions to this are when a particular species is endangered, or if it cannot be kept without undue stress in laboratories. Also when the way in which the animal is obtained is inhumane and unacceptable, e.g. the capture of chimpanzees where several females die to obtain one live baby chimpanzee, and when the way in which the animals are kept in laboratories is unacceptable (see FRAME/CRAE (1987) for a fuller explanation with regard to primates).

Genetic variables and strain

In any scientific experiment there are two major components contributing to biological variation: genetic and environmental (see Weisbroth 1984). It is possible to reduce genetic variation by the use of inbred strains which are produced by brother–sister or parent–offspring matings for at least 20 generations, after which time they are greater than 99% homozygous. The use of inbred strains should reduce the variance attributable to any genetic component in the experimental results, but some argue that the sensitivity of such a model and its relevance to humans may also be reduced because of this restricted genetic diversity. Humans, after all, are outbred animals, and one has to define carefully the genetic parameters one wishes to fix (or vary) before coming to a decision.

Other types of genetically defined animals are commonly used. In immunological research, specific crossbred animals are produced, often from inbred strains, so that the offspring have a very precise and defined genetic difference, e.g. at a single specific histocompatibility locus (see Festing 1979). Furthermore, various genetic mutants have occurred over the years and been selected so that the mutation can be produced in a predictable manner (usually following Mendel's laws), and these animals can be reliable models of human and animal diseases if they have a common genetic origin (Bulfield 1980). The production of transgenic animals, where recombinant deoxyribonucleic acid (DNA) is inserted into the genome, will increase our choice of experimental models and hopefully could increase the validity and reliability of the results obtained (in their relevance to humans, i.e. produce better models), should reduce the numbers of animals used, and should increase our understanding of gene expression and regulation, and associated defects. In the future, transgenic animals may be available that are able to breed true for a particular genetic defect. It has to be mentioned that transgenic animals also are the subject of some concern as the transgene (or rather the point at which the transgene is inserted) may produce adverse affects.

Common inbred strains are available from many commercial suppliers as are some other mutant strains mentioned above. The less-common inbred strains may only be available from another research laboratory, and consulting the *International Index of Laboratory Animals* (1987) or a literature search may help to determine a source. By and large, the commercial supplier will charge more than will a research colleague, but the animals will often be of better health status (see below).

Genetic monitoring

It is always possible that during the production of an inbred strain there may have been a mix-up in the breeding (I believe this happens extremely rarely but when it occurs it can have considerable implications on the research), and so confirmation that the animals are the designated strain should always be requested. Genetic monitoring will routinely be carried out by the better animal suppliers. Monitoring can be carried out in various ways: by phenotype (often coat colour or type), by testing histocompatibility through tissue (skin) grafting, by serology (immune reactions with antisera), by blood enzyme/chemical profiles, by analysis of the shape of the mandible, and by DNA or genetic fingerprinting. While still at the development stage it is likely that the latter technique will be able to identify positively a particular strain. The others will give a far lower probability of how unlikely, or likely, an animal is of that strain.

Environment

Environmental conditions can vary enormously between laboratories owing to differences in bedding, food, temperature, relative humidity, light rhythms and intensity, noise (including ultrasound levels), general husbandry of the animals (e.g. cleaning of cages), build-up of noxious gases (such as ammonia) and technician skills (see Clough 1981). The advantage of carefully controlling these variables is that it permits variation of only one factor at a time, e.g. dose rate of a drug, and it is more likely that the work will be valid and reproducible in another laboratory. Furthermore, the environmental parameters should be recorded when writing up experiments for publication (Working Committee for the Biological Characterisation of Laboratory Animals 1985). It is likely that the guidelines produced by the Royal Society and the Universities Federation for Animal Welfare (1987) will become a useful reference for environmental controls and for cage sizes and other matters of husbandry, and indeed have formed the basis for a Code of Practice now issued by the Government under the Animals (Scientific Procedures) Act 1986. The UFAW handbook is an invaluable source of information on the care and husbandry of many laboratory animal species (UFAW 1987).

The animal house staff will be aware of many of the common pitfalls that can affect the research purely as a result of differing husbandry conditions. They are already likely to be carrying out monitoring of the environment, e.g. temperature, relative humidity, air flow rates. It is important that the environment is controlled and records kept as this helps to ensure that comparable work can be done in another laboratory. Seemingly trivial differences may be responsible for not being able to do so, leading to unnecessary animal use; that is why full details should always be recorded in a scientific paper.

Diet and bedding

A considerable amount of work goes into the quality control of the food and bedding given to laboratory animals. Diets are now carefully analysed, as in the past they have been shown to be deficient in vital substances (e.g. vitamin C for guinea-pigs) or contaminated by unwholesome raw materials, during processing or storage (e.g. moulds, toxic metals or agricultural chemicals; see Clarke *et al.* (1977)). Meeting good laboratory practice standards (as laid down by the United States Food and Drugs Administration) helps to ensure that the diet will be wholesome and adequate, and that the bedding will not contain materials that may cause unwanted effects during the experiment. For example, sawdust bedding can cause artefacts in experimental results when made from redwoods which contain volatile oils, as these induce microsomal enzymes in the liver.

Semi-synthetic diets are commonly used and contain part-natural and part-defined chemicals; they have to be supplemented with vitamins and minerals to prevent deficiencies occurring. Totally chemically defined diets are difficult to make and expensive. It is always useful to consult the animal house staff and a professional nutritionist (most of the diet suppliers will be able to provide you with names and addresses) whenever the diet needs to be varied. This will ensure that the effect looked for is adequately controlled in the 'control' diet (for example by matching metabolisable energy levels, carbohydrate, protein and fat levels, etc.) and that no inadvertent vitamin, mineral, fatty acid or amino acid deficiencies are likely to occur.

Disease

Quality control (or quality assurance) of animals is a term used to encompass two aspects: genetic monitoring to ensure that the animal is of the strain required (particularly important when dealing with inbred strains, above), and health status.

Determining the health status of an animal means being aware of the organisms it carries, or with which it has been in contact. This is important as some diseases may affect the research (Weisbroth 1984), or the animal may succumb to a disease in the middle of an experiment, owing to the stress produced by the scientific procedure. We know also that many of the viral diseases in rodents have immunoregulatory effects—Sendai virus can suppress immunity, whereas under certain conditions mouse hepatitis virus can increase their resistance to infection, possibly through raising their immunocompetence.

There are several recognised classes of animals with regard to their health status: 'germ-free' with no microbiological fauna or flora; 'gnotobiotic' with a defined and deliberately administered micro-biological flora; free from specified pathogens (so-called 'SPF' animals); and finally 'conventional' which are normally free from diseases communicable to humans (zoonoses) and other diseases that are communicable to other animals but fairly easy to eradicate. The first three groups are normally kept behind a barrier to prevent contamination whereas the last group are kept conventionally with relatively few restrictions on access, diet, water, bedding, etc. Many experiments require only healthy animals, not those that are free from contact with these other infections. However, at least a conscious decision should be made to choose a particular health status of animal with due consideration for the science, rather than using what is readily available.

Microbiological screening is likely to have been carried out before the animals arrive at the animal house, as many of the commercial suppliers have joined a registration scheme (members of the Laboratory Animal Breeders Association Accreditation Scheme will supply details of screening programmes on request) to ensure that they supply animals of a defined quality. The scheme checks for organisms known to be pathogenic for animals and humans at regular intervals, and two categories are defined: barrier and non-barrier (with the barrier-maintained animals being of specified pathogen-free (SPF) status). For species that are not normally screened, or for which screening data are not available, then information should always be obtained to ensure that the animals are not carrying a disease that could infect people, or that might prejudice the general health of the individual animal or the colony: otherwise an infection might be brought into the animal house which could ruin valuable experiments and result in a waste of animals' lives.

A potential disadvantage, however, is buying-in animals of a higher health status (e.g. SPF) than the ones which they will come into contact with in the local animal house (e.g. conventional), i.e. buying-in animals that are 'clean' does not mean that they will remain clean in the user's animal house. Extra precautions may have to be taken to ensure that the health status of the animals is maintained, e.g. by keeping them in isolators, or behind a barrier.

Finally, although most animals used for research are purpose bred by commercial suppliers, there may be a reason why wild-caught animals have to be used, e.g. in zoological research. Confinement in a laboratory of these animals should not be undertaken lightly. Not only are the animals of an undefined health status, with all those attendant risks, but there are also the additional stress-related problems associated with confining a wild animal. The consequence of this stress is likely to add variability to the results and raises ethical questions as to whether the research should be carried out.

Obtain statistical advice

A statistician will advise on the design of the experiments and the number of animals needed in each group, and this should be done *before* carrying out any experiments. It may be useful to consult again after some preliminary results have been obtained, as the standard errors may be so large or so small that the group size may have to be altered. An introduction to statistics and the use of animals is given by Geller (1983) and Lovell (1985).

Refinement of experimental technique

Whatever model has been chosen there is always the possibility that the actual methods to be employed and the technical interferences can be modified so that the animal will suffer fewer adverse effects. This may involve, for example, choosing an anaesthetic that is non-irritant or with wide safety margins to ensure effective analgesia during an operation; cannulating an animal rather than taking repeated blood samples by venepuncture; or choosing injection sites and volumes that are compatible with the anatomical and physiological size of the animal. When operating on an animal, it should not be tied down so that it cannot show a response if the anaesthetic becomes too light or so that the animal subsequently experiences some discomfort or paralysis owing to an over-abduction of the limbs. Sutures should be placed so that they are not removed by the animal owing to the irritation associated with infection, or over-tightening, thus leaving an open wound (even if this is just for a few hours—animals live in close contact with their faeces and urine, which increases the chance for such infections).

Laboratory animal anaesthesia has changed considerably in the last few years and there are now many effective methods of anaesthetising even small mammals safely and reliably, thus obviating the need to use some of the older anaesthetics such as barbiturates, ether, chloroform, urethane and chloral hydrate (though some of these can have a useful role in terminal anaesthesia). Licensees should be encouraged to consult with the named veterinary surgeon on this matter as well as referring to standard texts (Green 1979, Flecknell 1987).

Raising antibodies in rabbits and mice is an ideal candidate for refining experimental protocols in order to avoid the large granulomata and suppurating abscesses that can form after using Freund's complete adjuvant. Choosing the site of injection, adjusting the volume and even modifying or eliminating the adjuvant can all improve the lot of the animal without loss of the scientific objective (see Amyx 1987).

The more conscientious researcher might like to pose the question: 'What would I want to be refined in the experimental protocol *before* that work was carried out on myself?' For example, if electric shocks are to be given as an aversive stimulus, is the current set at a level which is consistent with a response or is it set far above that; or better still why not reinforce by reward? What aftercare might you consider to be required—what frequency of checking/visiting? Is the anaesthetic to be used the most humane or that which is most convenient? While this may seem anthropomorphic, it does give some sort of basis on which to make the multiple technical choices required by experimental work.

LICENCES AND CERTIFICATES FOR ANIMAL EXPERIMENTS

Having decided on the factors discussed above and selected an animal model, a licence to carry out the work will almost certainly be required unless the experiment simply involves euthanasia by a straightforward method (i.e. those listed on Schedule 1 to the Animals (Scientific Procedures) Act 1986) such as an overdose of anaesthetic. Licences are issued under the Animals (Scientific Procedures) Act 1986 and it is worth mentioning the principles involved, as the licensing system forms the cornerstone of the Act. There are three types of licence:

● To approve and register the animal house facilities (a certificate of designation).

● To approve the programme of research work (a project licence).
● A certificate of competence for those carrying out scientific procedures on animals (a personal licence).

A new researcher may be in the position of not having held a licence before, and the work may be sufficiently novel so that a project licence (the licence which allows a programme of research work) may have to be modified in order to cover the work proposed. The following illustrates how the new system helps evaluate and minimise any adverse effects on the animals that may be produced, and how the facilities are scrutinised to ensure that any animals kept for research purposes are maintained in good health and conditions.

Certificate of designation

This licenses the animal facilities as a designated scientific procedure establishment (possibly also as a breeding and supplying establishment if it supplies animals to other scientific procedure establishments) and requires that it meets standards for the care and husbandry of the animals. As a requirement of this certificate, the certificate holder (normally somebody with considerable authority, such as the registrar of a university, member of the board of a commercial company, or a director of research) has to name two persons, a veterinary surgeon and a person in day-to-day care. Both these persons have a statutory duty to look after any animal that may give rise to concern. In effect they are the 'animal's friends', and it is part of their job to put, as it were, the animal's side of the case. They will help in interpreting the severity banding attached to the project licence (see below).

The balance of interests (see essay by Hampson in this volume) between science and the animals would seem to be an inevitable conflict that runs throughout animal experimentation. While good animal welfare, by and large, will lead to good science, there will be times when the animals' welfare will have to be compromised in some way in order for the scientifc objectives to be achieved. On these occasions, the presence of these named persons will help ensure that the balance of interests is maintained.

Welfare provisions in the personal licence

This licence is in effect a certificate of competence which helps ensure that those actually carrying out scientific interferences on animals have the necessary ability and manual dexterity, or are likely to gain such expertise to carry out these procedures humanely after sufficient training and experience. Much can be done to improve the welfare of animals and the quality of the research by reducing the stress caused by simple procedures, such as handling the animals carefully and considerately—even by not being afraid of them! Understandably, new (and even old) researchers may be nervous of handling a strange species, e.g. a rat or mouse, for the first time (the handling of strains of rats and mice can vary enormously), but acquiring such confidence will be of great benefit to researchers and their work and it is an area where animal distress and discomfort can easily be reduced. The animal house staff will be in a good position to help by choosing a strain that is quiet and docile (DB or Sprague Dawley) or an animal that is used to being handled—it may even be a 'pet'. It is also essential for the user to learn something about how the animals are kept.

Many of the animal house staff will have been trained and have qualifications from the Institute of Animal Technology (IAT) or the British Technician Education Council (BTEC). The IAT has three levels starting at Intermediate (basic animal biology, handling and husbandry), progressing to Associate and then on to Fellowship (senior technicians with considerable skill and experience). BTEC has equivalent grades with the Ordinary level being approximately equivalent to Associate and the Higher to the Fellowship. Some of the technicians, regardless of qualifications, will be Registered Animal Technicians, indicating they have an acceptable level of experience as animal technicians and are aware of the professional standards that are required as animal caretakers and their responsibilities for experimental animals. Researchers rely heavily on these technicians to look after the animals and to inform them whenever the animals are unexpectedly unwell.

Some animal houses have staff who carry out scientific techniques on behalf of the researchers, that is, they hold personal licences for a range of simple techniques such as injections and withdrawal of body fluids. This may be a way in which the welfare of laboratory animals can be improved, as these technicians will develop considerable expertise in some techniques as they will be carrying them out more frequently than the research worker who has a very limited number of animals. Furthermore, the animal technicians are likely to have considerable expertise in handling the animals and may even have gained their confidence because they look after them. Researchers should perhaps view the animal house staff as a valuable source of advice and expertise, especially if they are not confident of carrying out the work properly themselves. Researchers must always remember that the animals are their responsibility—those animals would not be used were it not for the research interests of the personal licensee.

The techniques that are to be used during an experiment may have to be learned, and advice can be sought from the research supervisor, research colleagues and the named persons. The person in charge of day-to-day care may know of other researchers using that technique or who have used it in the past, or a laboratory animal veterinarian who has specialised in laboratory animal science and medicine may be particularly knowledgeable and helpful. Similarly the Home Office Inspector, with a broad range of experience of different laboratories, may be willing to help make the necessary introductions. While others will only be able to demonstrate how to carry out a scientific procedure during the course of an experiment (it is not permitted to use an animal purely to gain or demonstrate manual skills—the only exception being for microsurgery), it will at least be helpful to see it being done. There may be some references to read and the potential licensee should be familiar with the anatomy and have practised on dead animals where this is possible. It is, of course, impossible to carry out some techniques on dead animals even moderately realistically, e.g. blood pressure monitoring and anaesthesia, but at least it is a start and something will be achieved even if it is only learning about the texture and manipulation of tissues (providing freshly killed animals are used). The person in charge of the local animal facilities may be able to help by providing recently killed animals which have been used in another experiment or are being culled from breeding programmes. Training courses are available at some institutions and videos are useful, but they will not permit practising on living animals. However, guidance on basic methodology, such as suturing or simple surgical approaches, can be learned and wasting animals through faulty technique may be avoided. If there is nobody carrying out the relevant techniques in the institution, then arrangements will have to be made to visit another laboratory to see them in operation.

Conditions and limitations attached to personal licences

Attached to a personal licence are many limitations and conditions. Three groups can be identified.

The first are 'standard limitations', breaches of which constitute an offence under the Animals (Scientific Procedures) Act 1986 and for which you can incur a heavy fine and be imprisoned. These include carrying out a regulated procedure not allowed on the personal licence, or on a species of animal not specified on the licence, or not working in those places specified in the licence, or not being covered by a project licence.

The second group are 'standard conditions', breaches of which do not constitute a criminal offence, but great importance is attached to their observation. Penalties can include revocation of the licence and hence research with animals. Responsibilities include:

● To label clearly all cages (so that an Inspector, and others, can identify the personal licensee responsible for that animal, the research project involved and the procedures being carried out).

- To kill any animal that is suffering at the end of an experiment, or that is in severe pain or severe distress regardless of whether the experiment has been completed or not.
- To take effective precautions to reduce suffering to a minimum level consistent with the aims of the scientific procedure (e.g. by using an analgesic or other alleviative treatment).
- To adhere to the conditions attached to the project licence, particularly with respect to the severity banding on the project licence (see below), and to consult with the project licence holder whenever the banding is exceeded.
- To ensure that the animals will be cared for in his/her absence (e.g. weekends and holidays) and to obtain veterinary advice whenever necessary.
- To adhere to the supervisor's requirements.
- To ensure that any aftercare of the animal meets an adequate standard.

There are also other conditions attached to a personal licence involving re-use of animals, neuromuscular blocking agents and the disposal of animals after an experiment has been completed (it is now possible to re-home experimental animals providing they are suffering no adverse effects from the research procedures).

The third group of conditions are not standard and are those that Home Office Inspectors may decide to add at their discretion. Breaches of these are treated in a manner similar to the standard conditions.

Welfare provisions in a project licence

The project licence must provide the Home Secretary and his advisers with all the necessary information to carry out a cost–benefit analysis on the proposed research programme. There should, however, be no difficulty in providing much of the necessary information, as the experiments will have been planned thoroughly through research and consultation, and often after peer review for grant application (this will help to avoid any duplication of research effort). Considerable thought will have been required to provide such detail (some of which is predictive) but the licence when granted will specify closely the limits of animal suffering allowed, methods of alleviation and specified end-points, thus helping to meet the requirement on the personal licence to reduce pain, distress and discomfort to the minimum level consistent with attaining the scientific objective.

The project licence holder is responsible for the use of animals, the personal licensees working on the project, and for observing the conditions attached to the licence. There should nearly always be a deputy project licence holder to act in the (unexpected) absence of the project licence holder. The deputy will normally be a personal licence holder (and this would be so in all instances if the project licence holder was not a personal licence holder). A condition attached to every personal licence ensures that if an experiment exceeds the allocated severity banding then the project licence holder must be informed (hence the importance for the personal licence holders to arrange for cover during any absence, or leave telephone numbers so that they may be contacted easily). Therefore, either both the project and personal licence holders should be contactable or firm alternative arrangements made—possibly with one of the named persons—so that appropriate action can always be taken. This will involve discussions between the named persons, the animal technicians and the licensees to highlight those signs an animal could show which would indicate that it had exceeded a severity limit, and may lead on to alleviative treatment or termination of the experiment. It is to ensure that requirements of the project licence are met that a copy should be kept in the animal facility or be easily available to the named persons.

If the research is to be carried out with someone at another designated scientific procedure establishment, then there will also have to be a deputy project licence holder at the second establishment to uphold the conditions and limitations in the project licence. In zoological studies involving the trapping of wild animals, the precise location may not be known. However, the Home Office require that a rough location be given (such as the Ordnance Survey grid reference) as well as notice of the intended work so that an Inspector has the opportunity to see the work in progress.

Project licences are only valid for a maximum of five years. This means that the cost–benefit analysis has to be regularly reassessed and any advancement in the treatment of animals or the refinement of any technique can be incorporated into a new project licence at that time. The Secretary of State can also refuse to grant, or not renew, a licence without recourse to further legislation for certain types of research work (on the recommendation of Parliament or the Animal Procedures Committee).

It is recognised that there are many legitimate reasons for carrying out research on animals and these reasons have to be specified on the licence (a broad range of permitted purposes is given in the application form notes and one or more of these have to be selected to indicate the reasons for the research). But some research techniques are of particular interest or concern (such as those involving trauma, scalds and burns, the special senses or central nervous system) and these have to be indicated on the project licence. Furthermore, applications for licences in some areas automatically have to be referred to the Animal

Procedures Committee (gaining manual skills in microsurgery, tobacco and cosmetic testing).

The key questions in the project licence in relation to the welfare of the animals are those which form the basis for the cost–benefit analysis. In the application the background, objectives and potential benefit of the research are given, as well as the practical ways in which these objectives are to be achieved (i.e. an outline of the research plan). The potential benefits may be in applied medicine for humans or animals or simply an increase in our scientific knowledge. However, it is likely that higher severity bandings will be permitted the greater is the potential benefit to the relief of human suffering.

An estimate of the number of animals of each species to be used including stages of development is requested (foetal forms are now protected from 50% of the way through gestation or incubation) and this will help reassure the Inspector that excessive numbers will not be used.

Recognition and alleviation of adverse effects as part of the project licence

It is worth pointing out straight away that pain, distress and discomfort are usually completely unintended side-effects of the experiment, and serve, if anything, to confound the results. (An exception is the testing of analgesic drugs.) Consequently the more they can be minimised, the better will be the research. For example, due consideration should be given to monitoring post-operative pain and the routine use of analgesics—even in rats. I use the example of rats deliberately as they can be regarded as vermin and, therefore, unworthy of such consideration—a public conception rather than a scientific one, as there is no biological evidence that they suffer pain any less than any other species of mammal, and so we should consider the use of analgesics in the rat just as we might do with dogs or the human species.

It is the recognition of these adverse effects that causes concern. While some would argue that one should wait until it can be *proved* that animals feel pain, and we cannot be sure that animals are in pain, it seems to me that we should give the animal the benefit of the doubt. There are undoubtedly differences between animals and people, but the maxim put forward by Professor Patrick Wall that 'painful conditions in man should be assumed to be painful in animals' is a very good starting point (see also Zimmerman 1983). Similarly any treatment for the pain could in the first instance be based on that in humans. After all, we use animals as models for pain in humans, and much of our basic knowledge of pain pathways and perception comes from experimental animal work; it would, therefore, seem equally logical to reverse the argument. Perhaps the most important difference is the way in which animals show pain. They cannot communicate (to *us*) their feelings, but no more can human babies, and yet many mothers, doctors and nurses would not deny that they feel pain, distress and discomfort.

The recognition of these adverse effects, then, is the nub of the problem and this differs between species. Consider also the differences between types of pain—compare chronic pain (e.g. a long-lasting moderate pain as in arthritis) with acute pain (e.g. trapping a paw or foot in a door). Several papers have been written on the recognition of these adverse effects and on their alleviation (Morton and Griffiths 1985, Duncan and Molony 1986, Morton 1986, Association of Veterinary Teachers and Research Workers 1986, Flecknell 1985, American Veterinary Medical Association 1987), and it may be helpful to read some of these before starting an experiment.

One of the most important features of the project licensing system is that a limit will be placed on the amount of suffering that is permitted, and this is achieved through the severity limit attached to each scientific procedure (there may be several procedures within each project and each one will have a specific severity band attached to it). Three bands of severity are ascribed: mild, moderate and substantial. Severe pain or severe distress is also a recognised condition, but it is not permitted as the animal must be put down, or the condition alleviated—this is a condition on the personal licence and is often known as the termination condition.

The adverse effects of a procedure on an animal and their likely incidence requires considerable thought because they are *predicted* adverse effects. If one carries out an operation then certain events may happen, e.g. post-operative infection, pain, haemorrhage, tissue rejection. It is important to consider not only the likelihood of adverse effects, but also their alleviation in order to avoid unnecessary animal suffering. In order to reassure oneself that the best advice has been taken, and that animal welfare will be best protected during and after any procedure, all those who will have a role to play with those animals under experiment should be consulted (the licensees, the named persons and experienced research scientists). After consulting with colleagues, perhaps a shortlist of likely signs of adverse effects could be drawn up, together with prescribed courses of action, so that all those involved with the care of the animals would be on the lookout and could take remedial action at an early stage. The experience gained from the first set of animals will prove to be a valuable source of information and they should be monitored extremely closely by all concerned (see tables in Morton 1987).

As well as alleviating any side-effects it is obviously better practice to consider how to prevent them occurring in the first place e.g. by operating with full aseptic precautions, causing as little tissue damage as possible, providing good anaesthesia, keeping the animal warm during the operation, giving post-operative fluids and attending to its normal physiological requirements during the experiments. The named veterinary surgeon should be able to help on these aspects. Always observe animals carefully after any procedure, particularly if it is one that is novel. Licensees must ensure that their manual dexterity is good enough to perform the technique precisely, efficiently, accurately and repeatably, and to put down any animal in which the technical side has been deficient.

HUMANE ENDPOINTS

It is my view that no animal should ever deliberately be left to die in an experiment without the severest scrutiny of the research protocol. It is in this context that humane endpoints are important—these may be pre-lethal or, better still, pre-painful. Lethality tests are by definition usually unpleasant and involve a considerable amount of suffering (exceptions can always be found, e.g. overdose of anaesthetics, but these do not detract from the overall argument). In toxicity tests pre-lethal and pre-painful endpoints should be sought whenever possible (greater justification is likely to be required for a project licence for work of this sort) especially as, from the scientific point of view, much valuable information may be lost if the animal has been dead for several hours. It may be possible to determine such humane endpoints of sufficient reliability that comparative toxic or virulence levels can be established (British Toxicology Society 1984, Fielder *et al.* 1987, van den Heuvel *et al.* 1987) which are acceptable to regulatory authorities. In some cases, those in which the test is expected to be completed in a few hours or a few days, it may be possible to carry out the test on animals that are terminally anaesthetised or which have been decerebrated (and are therefore unable to perceive pain), but in this latter group artificial feeding will be required.

Examples of pre-lethal endpoints could be elevation or lowering of blood glucose to a particular level as an indication that a pancreas transplant had failed or that a standard dose of insulin was effective. Similarly, signs of disease caused while a vaccine is undergoing a batch production test, which would make it unacceptable for marketing, should become the endpoint rather than death (the same could be applied to the 'standard positive control'). In cancer research, after a tumour passage or the deliberate induction of malignant growths with a carcinogenic agent, there is very little point in letting these tumours grow to such a size that they incapacitate the animal, or become ulcerated with secondary infection; in fact it is often to the detriment of the research when this occurs. Survival times as a measure of treatment efficacy are crude and inhumane and can be replaced by humane alternatives (Hewitt 1981).

Re-use of animals as a means of reducing the number used

Any re-use of animals from one project to another, as well as from one procedure to another within a single project, will have to be fully detailed and justified in the project licence. In any re-use from one *project* to another, when the animal has been given a general anaesthetic in the first use, the second use can only be under terminal anaesthesia and only then if it is fit and healthy after the first use and not suffering any adverse effects (this has to be certified by a veterinary surgeon and also included in the project licence). Furthermore, the time between the first and second re-use must not be over-long, nor is there any question of animals being returned to a breeding or supplying establishment for re-issue. While the occasions when an animal can be re-used are likely to be very limited, it will reduce the number used, particularly of those species that cause most concern (primates, dogs and cats). For example, whereas a dog with a successful kidney transplant may be put down at the end of 100 days, as a period arbitrarily taken as indicative of success, the animal could be used under terminal anaesthesia for a second unrelated experiment on blood flow studies instead of having to use a second dog.

Neuromuscular blocking agents

Neuromuscular blocking agents cause concern to all who use them, as they have obviously potential harmful effects (blocking a conscious response to pain by paralysing the muscles). Details and justification of any proposed use are required by the Home Office in the licence application and there are useful guidelines available from the Physiological Society and the Home Office.

FUTURE DEVELOPMENTS IN THE WELFARE OF LABORATORY ANIMALS

Much research is being carried out in caging design and the way in which we group animals—what effect does housing animals singly have on those species with distinct natural social interactive behaviours, e.g. dogs and primates? This must be welcomed, both for humanitarian and scientific reasons. This is a broad subject but suffice it to say that group size, methods of handling animals, variation in the environment and technical skills have all been shown to affect research results. We should be aware of what influences animal behaviour and of the needs of the animals, and then plan experiments accordingly.

Other areas in science worthy of note are the increasing ways in which non-invasive methods of monitoring human health, and therefore animal health and function, are being developed. For example, nuclear magnetic resonance (NMR) and computerised axial tomography (CAT) scanning will enable the early detection of tumour growth, avoiding anaesthesia and exploratory surgery that otherwise would be required to detect a deep tumour. Developments in monoclonal antibody raising (especially those that avoid the use of ascitic animals), transgenic animals and synthetic compounds may all help answer scientific questions more accurately and reduce the number of animals used in research. Similarly, advances in alternatives such as computer modelling, cell and organ culture, validation of *in vitro* techniques and interactive videos for teaching will all lead to a reduction of animal use (see essay by Stephens in this volume).

Euthanasia

The commonest procedure carried out on animals is humane killing or euthanasia. Understandably, it is perhaps the animal house staff and the veterinary surgeon who will have the most expertise in the techniques involved and these persons will give useful advice. Schedule 1 to the Animals (Scientific Procedures) Act 1986 sets out those techniques which require least skill and can be carried out without a personal licence. Careful consideration should always be given to the method of euthanasia, as frequently the reasons for choosing a method are based on supposition rather than fact, with no reference to science or animal welfare. Physical methods have the advantage in that there is no 'contamination' by extraneous chemicals and they are quick (though restraint can be distressful), but have the disadvantage that they can be distasteful to the user, and there is room for error, with untoward consequences. Contamination by an overdose of anaesthetic more often than not has little or no effect on the research and is probably the method of choice (from the animal's point of view) in most circumstances.

CONCLUSIONS

In this essay I have tried to indicate where I consider animal welfare can be improved and at the same time where science can benefit, as good animal welfare and good science often go hand-in-hand. Where they do not, then justification is required to be sure that the basis for increased animal suffering is well founded. The Animals (Scientific Procedures) Act 1986 provides the basic framework, partly through its licensing system, to make certain that animal welfare is given prime consideration without affecting the legitimate aims of the research. The Act will make for greater standardisation throughout the UK, more justification for the use of animals, increased exchange of information and better supervision of animal-based research. All of these in turn will lead to 'best practice becoming common practice'. Guidelines and codes of practice now issued and approved under the Act will also help in this regard.

It should be remembered that, regardless of laws, animal welfare is best served and protected by the individual. It is up to each research worker to consider the ethical issues involved and to choose techniques that will not only answer the scientific question but cause least distress to the animal. Serious consideration should be given to whether the knowledge is worth gaining—at the end of the day that is a personal decision (given that the research would be licensed under the Act). The researcher should bear in mind that not only are there moral responsibilities to the animal and the public, there is also a corporate responsibility to his or her colleagues and the scientific community, to uphold and engender high moral standards for animal research.

REFERENCES

American Veterinary Medical Association (1987). Colloquium on recognition and alleviation of pain and distress, *J. Am. Vet. Med.*, **191**, (10)

Amyx, H. L. (1987). Control of pain and distress in antibody production and infectious disease studies, *J. Am. Vet. Med.*, **191**, 1287–9

Association of Veterinary Teachers and Research Workers (1986). Guidelines for the recognition and assessment of pain in animals, *Vet. Rec.*, **118**, 334–8

Bentham, J. (1789). In Harrison, W. (ed.) (1967). *An Introduction to the Principles of Morals and Legislation*, Anchor Books, New York

British Toxicology Society Working Party on Toxicity (1984). A new approach to the classification of substances and preparations on the basis of their acute toxicity, *Hum. Toxicol.*, **3**, 85–92

Bulfield, G. (1980). Inherited metabolic disease in laboratory animals, *J. Inherit Metab. Dis.*, **3**, 133–43

Clarke, H. E., Coates, M. E., Eva, J. K., Ford, D. J., Milner, C. K., O'Donoghue, P. N., Scott, P. P. and Ward, R. J. (1977). *Dietary Standards for Laboratory Animals: Report of the Laboratory Animals Centre Diets Advisory Committee*, Medical Research Council, no. 11, London, pp. 1–28

Clough, G. (1981). Environmental effects on animals used in biomedical research, *Biol. Rev.*, **57**, 487–523

Dawkins, M. S. (1980). *Animal Suffering: The Science of Animal Welfare*, Chapman & Hall, London

Duncan, I. J. H. and Molony, V. (eds) (1986). *Assessing Pain in Farm Animals*, Commission of the European Communities, Luxembourg

Festing, M. F. W. (1979). *Inbred Strains in Biomedical Research*, Macmillan, London

Fielder, R. J., Gaunt, I. F., Rhodes, C., Sullivan, F. M. and Swanston, D. W. (1987). A hierarchical approach to the assessment of dermal and ocular irritancy: a report by the British Toxicology Society Working Party on Irritancy, *Hum. Toxicol.*, **6**, 269–78

Flecknell, P. A. (1985). In Fox, M. D. and Mickley, L. D. (eds), *Advances in Animal Welfare Science*, Martinus Nijhoff, The Hague, pp. 61–78

Flecknell, P. A. (1987). *Laboratory Animal Anesthesia*, Academic, San Diego

FRAME/CRAE (1987). *The Use of Non-Human Primates as Laboratory Animals in Great Britain*, FRAME, Nottingham

Gartner, K., Buttner, D., Dohler, K., Friedel, R., Lindena, J. and Trautschold, I. (1980). Stress response of rats to handling and experimental procedures, *Lab. Anim.*, **14**, 267–74

Geller, N. L. (1983). Statistical strategies for animal conservation, *Ann. NY Acad. Sci.*, **406**, 20–31

Green, C. J. (1979). *Animal Anaesthesia*, Laboratory Animals Ltd, London

Hampson, J. E. and Silcock, S. R. (1985). Pain in laboratory animals—the case for refinement. In Marsh, N. and Haywood, S. (eds), *Animal Experimentation: Improvements and Alternatives*, FRAME, Nottingham

Hewitt, H. B. (1981). The use of animals in experimental cancer research. In Sperlinger, D. (ed.), *Animals in Research: New Perspectives in Animal Experimentation*, John Wiley, Chichester

International Index of Laboratory Animals (1987). In Festing, M. F. W. (ed.), *Laboratory Animals Handbook*, no. 10, 5th edn, Laboratory Animals Ltd, London

Kaplan, J. R., Manuck, S. B., Clarkson, T. B., Lusson, F. M., Taub, D. M. and Miller, E. W. (1983). Social stress and atherosclerosis in normo-cholesterolemic monkeys, *Science*, **220**, 733–5

Lovell, D. P. (1985). Statistical approaches in toxicology: the use of factorial designs and multivariate methods, *Med. Inf. (Lond.)*, **10**, 143–51

Moberg, G. P. (1985). *Animal Stress*, American Physiological Society, Bethesda, MD

Morton, D. B. (1986) The recognition of pain, distress and discomfort in small laboratory mammals and its assessment. In Gibson, T. E. (ed.), *The Detection and Relief of Pain in Animals*, BVA Animal Welfare Foundation, London, pp. 64–74

Morton, D. B. (1987). Epilogue: summarisation of Colloquium highlights from an international perspective, *J. Am. Vet Med.*, **191**, 1292–6

Morton, D. B. and Griffiths, P. H. M. (1985). Guidelines on the recognition of pain, distress and discomfort in experimental animals and an hypothesis for assessment, *Vet. Rec.*, **116**, 431–6

Nerem, J. F., Levesque, M. J. and Cornhill, J. F. (1980). Social environment as a factor in diet-induced atherosclerosis, *Science*, **208**, 1475–6

Riley, V. (1981). Psychoneuroendocrine influences on immunocompetence and neoplasia, *Science*, **212**, 1100–9

Rowan, A. N. (1984). *Of Mice, Models and Men: A Critical Evaluation of Animal Research*, State University of New York Press, New York

Royal Society and Universities Federation for Animal Welfare (1987). *Guidelines on the Care and Housing of Laboratory Animals and their Use for Scientific Purposes*, 1, *Housing and Care*

Russell, W. M. S. and Burch, R. L. (1959). *The Principles of Humane Experimental Technique*, Methuen, London

Silcock, S. R. (1986). Refinement of experimental procedures, *ATLA*, **14**, 72–84

Smyth, D. H. (1978). *Alternatives to Animal Experiments*, Scolar, London

UFAW (Universities Federation for Animal Welfare) (1987). *The UFAW Handbook on the Care and Management of Laboratory Animals*, Poole, T. (ed.), Longman, Harlow

van den Heuvel, M. J., Dayan, A. D. and Shillaker, R. O. (1987). Evaluation of the BTS approach to the testing of substances and preparations for their acute toxicity, *Hum. Toxicol.*, **6**, 279–91

Weber, H. W. and van der Walt, J. J. (1973). Cardiomyopathy in crowded rabbits, *S. Afr. Med. J.*, **47**, 1591–5

Weisbroth, S. H. (1984). The impact of infectious disease on rodent and genetic studies, *Lab. Anim.*, **13**, 25–33

Working Committee for the Biological Characterisation of Laboratory Animals (1985). Guidelines for the specification of animals and husbandry methods when reporting the results of animal experiments, *Lab. Anim.*, **19**, 106–8

Zimmerman, M. (1983). Ethical guidelines for investigations of experimental pain in conscious animals, *Pain*, **16**, 109–10

Organisations and other sources of information

Canadian Council on Animal Care Resource, 100–151 Slater Street, Ottawa, Ontario, K1P 5H3, Canada

FRAME Toxicology Data Bank, FRAME, 34 Stoney Street, Nottingham, Notts NG1 1NB, UK

Institute of Laboratory Animal Resources (ILAR), National Research Council, 2101 Constitution Avenue, Washington, DC 20418, USA

Scientists' Center for Animal Welfare, 4805 St Elmo Avenue, Bethesda, MD 20814, USA

9. Plea for a Sensitive Science

Gill Langley

'A true science cannot possibly ignore the solid incontrovertible fact, that the practice of vivisection is revolting to the human conscience, even among the ordinary members of a not over-sensitive society. The so-called "science" . . . which deliberately overlooks this fact, and confines its view to the material aspects of the problem, is not science at all, but a one-sided assertion of the views which find favour with a particular class of men.'

Henry Salt (1892). *Animals' Rights Considered in Relation to Social Progress*, p. 97

INTRODUCTION

The use of animals in scientific experiments has been under fire from animal-welfare and anti-vivisection societies for more than a century. During that time, although there has been a steady decrease since 1976, the number of experiments conducted in Britain each year mushroomed from a few hundred to the present level of about three million. The Cruelty to Animals Act 1876, which effected some control of animal research, remained solidly in place for 111 years, despite the increasing difficulties experienced by the Home Office in applying it effectively to the nature and range of animal experiments and tests carried out in the late 20th century. Only when it became obvious that, as a result of political lobbying by the animal-welfare movement in the mid-1970s, new legislation was bound to be drafted, did the scientific community as a whole admit that a new law was needed to replace the old Act and begin to launch a damage-limitation exercise.

Russell and Burch, two scientists somewhat ahead of their time in their thinking, recommended in 1959 that efforts should be made to *replace* animals with non-sentient models, to *reduce* numbers of experiments and numbers of animals used and to *refine* procedures to cause less suffering (Russell and Burch 1959). Because of interest in developing cheaper or more reliable techniques, especially in commercial companies, some non-animal models were developed on a piecemeal, uncoordinated basis, but there was little sign of Russell and Burch's pleas being taken to heart and acted upon with any real motivation. This was provided more energetically by several anti-vivisection societies who, in the late 1960s, put their money where their beliefs were and established humane research charities to promote the development of non-animal techniques. Meanwhile, training courses in the care and handling of laboratory animals, and in the recognition of pain in different species, were few and far between, making a mockery of the law's much-vaunted control of pain in experimental animals. Little research into the physical and psychological needs of caged laboratory animals was initiated, even in the case of primates, whose behavioural and social complexities were beginning to be understood as a result of ethological studies. Techniques of anaesthesia and analgesia remained, and still remain, crude for many species used in research.

It was public outrage, following animal-welfare campaigns in the early 1970s, about the 'smoking beagles' research which resulted in the ending of the use of dogs for smoking experiments in Britain. Effective and professional campaigns against the LD50 and Draize eye tests during the 1980s drew scientists' attention to the unreliability and inhumanity of these tests, which had been used routinely since 1927 and 1944, respectively. The crucial role of the animal-welfare and anti-vivisection societies in highlighting the need for control, reform or proscription of animal experiments has been acknowledged in many places, for example in Clive Hollands' and Judith Hampson's essays in this volume, and in the journal *Science* (Holden 1987):

> Prodded by the animal welfare movement, major manufacturers of pharmaceuticals, pesticides, and household products have made significant advances in recent years toward the goal of reducing the number of animals used in toxicity testing. Alternative methods, such as cell and tissue culture and computer modelling, are increasingly being seen not just as good public relations but as desirable both economically and scientifically.

In this essay I want to explore in some detail the connection between the nature of scientific education and the hierarchies and prejudices that are found in the world of science, and the reluctance of researchers to tackle the ethical and welfare questions raised by animal experimentation. Of necessity I shall be referring to 'scientists' as though they are a homogeneous group; I am well aware that this is not so, but my remarks do apply to the majority of the scientific community and to the general state of the teaching and practice of science.

SCIENCE EDUCATION

From school level onwards, young scientists absorb the doctrines, prejudices and biases inherent in the prevailing practice of science. Senior researchers who are leaders in their fields write the key textbooks, append their names (although not always their effort) to numerous research papers and are also the lecturers who teach university students. Aspiring scientists are created, to a large extent, in their teachers' image, and will use similar parameters, concepts and practices in their turn, since a science education is not only the absorption of facts but also of attitudes (Morley 1978). This tendency for dogma to be perpetuated from one generation of researchers to another has important ramifications for animals in science.

One of the first doctrines a student of biology learns, even at school, is that it is perfectly correct to utilise other animals, alive or killed for the purpose, as scientific tools. The requirement that this use, if acceptable at all, should be made with sensitive concern for the welfare of each individual animal has not always been emphasised strongly enough. The initiation ceremony or 'first blooding' has inevitably been school dissection, and at university the student's attitude to animals is further modified. Confronted with a class laboratory festooned with decerebrate rabbits in my first year as a physiology student at Cambridge University in the early 1970s, my initial response was one of horror and revulsion. It seemed—and still seems—barbaric that nearly 300 students should use almost 150 rabbits to demonstrate known facts about the cardiovascular system described in any textbook. A student delegation elicited the response that it was important that we each have 'hands on' experience of living tissues to engender a proper appreciation of their delicate nature.

Curiosity, the pressure to conform and the knowledge that this 'experiment' could be set in the end-of-year exam conspired to overcome our initial, emotional (and justifiably so) reaction. For myself, this later gave way to self-disgust, as I brooded on the fact that, within a very short space of time, I had ceased to see the rabbit as a recently living creature but instead as a fascinating 'mammalian preparation'. This process of desensitisation to the uniqueness and irreplaceability of an individual animal and to its right to life allows a distance to be interpolated between *us* and *them*. Scientists dissociate themselves from the suffering which is inflicted on laboratory animals during experimentation, and from the massive loss of life caused by student classes in physiology, pharmacology, biochemistry and psychology at universities in Britain and, even more so, in the USA.

Desensitisation is an inescapable part of biology education and is, no doubt, a necessary form of self-protection, because for a scientist to empathise with laboratory animals could be emotionally devastating. That desensitisation occurs is sometimes denied. William Paton, then Professor of Pharmacology of Oxford University, in his book *Man and*

Mouse (Paton 1984, p. 149), claimed that desensitisation of surgeons and animal experimenters alike is merely a confusion in the minds of laypeople of:

> . . . the superficial appearance of lack of feeling with the actual exercise of professional skill and efficiency. . . . My own experience is that animal work makes the investigator *more* sensitive to animal needs, as he learns about their behaviour, their physiology, and gains experience in seemingly small but important matters about how to handle them.

Rather, the confusion is in Paton's mind. Surgeons operate in the knowledge that their efforts will benefit the individual patients under their scalpel, while animal experimenters cannot salve their consciences with any such belief, although they may well reassure themselves that their experiments could, one day and in some ill-defined way, benefit individuals of another species. Moreover, the idea that a sensitivity to animal needs is engendered by animal experimentation, as suggested by Paton, is heavily contradicted by the conditions in which, for example, laboratory primates find themselves. Jane Goodall, who has studied individual primates in their natural environment and has learned of their complex social relationships, their friendships, their fears and their joys (Goodall 1971), has contributed more to an understanding of these close relatives of ours than any sterile laboratory study. Goodall is now spearheading a campaign to improve the conditions under which primates are kept in laboratories, and has written (Goodall 1987):

> In most labs the prisoners cannot even lie with arms and legs outstretched, let alone walk, climb or swing. They are not let out to exercise. There is seldom anything for them to do other than eat, when food is brought. The caretakers are usually too busy to pay much attention to individual chimpanzees, over and above the daily maintenance routine. The cages are bleak and sterile, with bars above, bars below, bars on every side. There is no comfort in them, no bedding, despite the fact that the inmates, infected with human disease, will often feel sick and miserable.

Keeping sensitive, intelligent animals in conditions such as these would not, surely, be possible unless desensitisation occurs. In the same article Goodall wrote that the animal technicians, vets and scientists involved come to accept the cruelty by forcing themselves to believe that it is necessary to reduce human suffering: 'Some become hard and callous in the process.' Those staff who retain their compassion and empathy for animals often leave, unable to endure the suffering.

Kelly (1986) has suggested that students exposed to animal suffering may be desensitised or numbed to hurting living beings, and that by providing such experiences in the teaching laboratory, science educators may be fostering the tolerance and acceptance of inhumaneness. Alice Heim, a psychologist, has also given examples of the desensitisation of students and teachers, including her own experience teaching students in tutorials (Heim 1981).

An essential part of the desensitisation process is the debasement of language which features in reports of animal experiments. A dead cat becomes a 'mammalian preparation'; the chemical induction of pain and inflammation in the joints of a rabbit is the creation of an 'animal model' of arthritis; repeated electric shocks to the feet of mice are 'aversive stimuli' or 'negative reinforcement'; at the end of an experiment, a rat is 'sacrificed' (presumably on the altar of science). Such euphemisms, as well as the use of the passive mood in research papers, while conforming to the objectivity and precision required by scientists, also serve to distance them and their readers from the reality of the procedures they inflict and their effects on animals. A horrendous example of euphemism was provided by an American report (in a British journal) of experiments with a mouse 'oral radiation death syndrome model', where animals died from 'nutritional insufficiency' caused by 'diffuse mucosal damage in the oral cavity' (Grigsby and Maruyama 1981). The simple translation is that mice were irradiated to the head, causing widespread radiation injuries in the mouth, leading to death from starvation. The sanitised writing style is adopted by young scientists who see it as an essential prerequisite for publication in journals. I recently received a letter from a first-year PhD student whose research topic was the 'environmental enrichment' of captive cotton-top tamarin monkeys, 'specifically . . . the design of substrates facilitating motor activities in the cage'. She was referring to the provision of branches and other items as play and exercise objects.

The use of euphemisms with the purpose of insulating people from an emotional link with their victims has a long history in both human and animal relationships. The Japanese scientists and doctors who experimented on prisoners of war in the 1940s at Harbin, in Manchuria, referred to them as '*maruta*', which means 'logs of wood'. 'Sometimes there were no anaesthetics. They screamed and screamed. But we didn't regard the logs as human beings. They were lumps of meat on a chopping block', said Naoji Uezono, who watched such experiments (Whymant 1983). Reducing people to sub-human caricatures, such as 'krauts' or 'gooks', allows soldiers to distance themselves psychologically from their enemies. The same function is presumably served by talk, among farmers and game managers, of 'crops' of animals which need to be 'harvested'. A cloak of respectability is thrown over some socially or morally dubious activities by the use of euphemisms such as 'liquidation' instead of assassination, and 'purveyors of meat' instead of butchers. Such techniques foster compartmentalisation of thought and limit the full realisation of what is being done, and this is described clearly in the context of attitudes to animals by Cora Diamond (Diamond 1981).

The consequences for young scientists of failing to absorb the scientific doctrine of total objectivity in regard to other living beings were illustrated by Bernard Rollin in his book, *Animal Rights and Human Morality* (1981). He recounted the experience of a young American psychology graduate conducting experiments on rats (Rollin 1981, pp. 109–10):

> The experiment was over, and he was faced with the problem of what to do with the animals. He approached his adviser, who replied, 'Sacrifice them.' 'How?', asked my friend, assuming that the professor would produce a hypodermic needle and barbiturates. 'Like this,' replied the instructor, dashing the head of the rat on the side of the workbench, breaking its neck. (While this is not in fact a cruel way to kill a rat if done correctly, since cervical dislocation causes instant death, it is not easy to learn and is highly offensive to the uninitiated.) My friend, a kind man, was horrified and said so. The professor fixed him in a cold gaze and said 'What's the matter, Smith, are you soft? Maybe you're not cut out to be a psychologist!'

Natural sensitivity and respect for other animals in young biologists are all too often scorned and caricatured as squeamishness by senior researchers. I include here an example I have written about elsewhere (Langley 1981) of the mentality too often encountered in senior scientists. In 1978, a zoology graduate at Cambridge University, concerned by an excessive use of animals in teaching, set up a study group of students and others to consider ways of reducing and replacing animals in class practicals. He wrote to a number of course teachers and heads of departments, seeking their interest and support of the group's very moderate aims. One reply, received from a senior experimental physiologist at the Craik Laboratory, was of particular interest and read as follows:

> Thank you for your letter about misuse of animals in teaching. I have no idea to what misuse you refer nor how you can employ

alternatives to animals in teaching zoology. Consequently I have so poor a concept of the aims of your group that I have no strong urge to support it. However, I should be distressed to learn that cases exist here where animals are misused for teaching purposes, and I should consider it my duty to improve things if I could. But, of course, I should much rather not get involved in a subject so charged with ignorance and emotion.

In contrast to this arrogance and insensitivity was the reply of Horace Barlow, a professor in the Physiological Laboratory at Cambridge, who wrote that he was not sure how much animals were misused for teaching purposes nor how viable the alternatives were, but that it was worth looking into. He also wrote 'Another thing worth giving some attention to would be the attitudes about animal use that are instilled when they are used for teaching. That aspect may be more important than the number used.'

The campaigns initiated by organisations such as Animal Aid, the National Anti-Vivisection Society and the Royal Society for the Prevention of Cruelty to Animals against compulsory animal dissection in British schools have been well-supported by school pupils themselves, and have had considerable effect. A joint statement made by the Association for Science Education, the Institute of Biology and the Universities Federation for Animal Welfare (ASE/IOB/UFAW 1986) acknowledged the controversy. The statement suggested that animal dissection is an important method of enquiry and of obtaining first-hand experience, but condemned repetitious dissection, and teachers were exhorted to discuss the ethical, aesthetic, educational and scientific implications of dissection with their students. The British Examining Boards have responded to the campaign by moving towards making animal dissection an optional rather than a compulsory activity for 'A' level biologists, although some students are still faced with the prospect of losing marks if they refuse dissection.

In higher education a student's right to decline to take part in procedures on animals, without penalty, is included in the Violence-Free Science Student Charter, which has been adopted by more than 30 university and college student unions. Country-wide, the National Union of Students agreed at its 1987 conference to support students who refuse on moral grounds to do animal experiments. In the USA, the American Medical Student Association and the Physicians Committee for Responsible Medicine have promoted a similar resolution, which also requests that students should have access to information about the procurement and treatment of animals used in classwork. A reduction in the use of animals for teaching known facts is likely to follow the increasing sophistication and availability of interactive computer simulation programs, such as those produced by Sheffield City Polytechnic, and other alternatives to animal procedures.

It is possible to teach biology in a way which preserves, rather than destroys, a reverence for life. Shapiro (1987) mentioned two proposals which could minimise the detached and manipulative attitudes towards animals which students may easily acquire. Rollin (1981) described the introduction of courses on the ethical, philosophical and social implications of science and its use of animals, at some centres of veterinary and biomedical science eduction in the USA. These have substantially modified students' attitudes to animals; have resulted in improvements in animal welfare standards; and have enhanced the reputation of the colleges involved.

SCIENTIFIC CONFORMITY

The lay-person is still heavily influenced by the image of science as a dynamic, progressive, frontier-expanding activity, and of its practitioners as dedicated teams of rational, open-minded professionals whose motivation is highly moral and whose concern is to benefit humanity (Delacôte 1987). Very few scientists have been as honest in their appraisal of their motivations as has Professor Sam Shuster (1977):

> The scientist is not powered by the sentiment of the soap opera: he is driven by the twin motives of curiosity and reward, and, of these, curiosity is the most persuasive . . . Desire to cure disease is really a curiosity about the disease process or the sublimated motive of reward—because fame and fortune are the greater for applied research . . . Of course scientists wish humanity well and will occasionally even work for its well-being; but that well-being is not and never can be our true motive.

On the whole, researchers have done little publicly to dispel their glamorous and unrealistic image. The decisions of scientists, the story goes, are freer than those of average mortals from the taint of personal gain and status and they weigh up new ideas objectively and without inherent bias, engaging in constructive debate about them and adopting enthusiastically those which seem valid.

One stereotype suggests that science progresses within a framework characterised by sequential steps: by the application of inductive logic to a set of discovered facts, leading to the framing of a hypothesis, which is then modified in the light of further specific experimentation. This rigorous approach is complemented by the creative stereotype (exemplified historically by Kekulé's discovery of the structure of the benzene ring) whose use of intuition and inspiration allows leaps of insight—described as the 'romantic view' of science by the late Peter Medawar (Medawar 1969). The intuitive scientist is on a par with the poet, whose freedom to create must be untrammelled by wordly concerns such as the demands of society or the need for accountability. There are, as Rowan (1984, p. 16) says, elements of truth in both stereotypes: but as models of the scientific mind they are simplistic and adulatory, since researchers' thoughts and actions are as susceptible to influences such as personal gain, status and an education which instils conformity, as any other group of people.

As young scientists advance through the education system, course options such as the humanities, social studies, ethics and the philosophy of science become restricted because of the increasing need to specialise. As well as losing the opportunity to see science in a broader context, this has the effect of drawing them deeper and deeper into that close-knit community wryly termed the 'scientific mafia' by one science journalist. Obtaining a PhD admits researchers into a specialised, high-status clique, where their behaviour reflects directly not only on themselves but also on others in the scientific community. Membership of the professional societies beckons, promising even closer admittance to the in-group and providing even greater pressures to conform. Indeed, the Physiological Society in the UK was established in 1876 in direct response to the embryonic anti-vivisection movement. The pressure to conform can be very strong indeed, even though it contradicts the apparently open-minded, objective notion of the practice of science. This is not to say that debate or reform is not possible within the scientific community, but only that it is restricted to issues within certain narrow confines and can be raised through only a few accepted channels. The examples I gave earlier of young scientists' concern for laboratory animal welfare—expressed spontaneously in one case, and through the formation of a (moderate) pressure group in the other—being immediately stamped upon by some senior researchers, illustrate the process of conditioning which takes place.

The opprobrium which falls on *senior* scientists who break rank is all the more severe. In the 1970s Harold Hewitt was a senior cancer researcher and head of the Tumour Radiobiology Section of the Cancer Research Campaign's Gray Laboratory in Northwood, Middlesex, and at the peak of his career. He had published several key research papers and had received an international award for his work on the application of radiation to medicine. Hewitt had been responsible for setting up and managing the animal house at the unit, whose standards of hygiene and animal care were acknowledged to be of the highest. In 1970 a new Director was appointed, a physicist with little experience of handling animals. Hewitt's chief animal technician, Angela Walder, became increasingly unpopular with the new Director and his staff, because of her outspoken stand on animal-welfare issues. She complained that researchers went on holiday without leaving clear directions regarding experimental animals, and that the metal ear-tags used by some scientists at the laboratory to identify rodents caused them distress.

In 1975 Walder was accused of deliberately supplying mice of the wrong strain to a scientist, 15 months previously. An internal enquiry exonerated her, but conflict between the Director and the animal house staff supported by Hewitt escalated further when two junior technicians were sacked. In 1976, designated Animal Welfare Year, Hewitt spoke to an *Observer* journalist of his concern about laboratory animal welfare in general, instancing his own experience but asking that the Gray Laboratory was not named to avoid attracting adverse publicity. After the article was published, the laboratory Director asked Hewitt if he was the source, which Hewitt admitted.

From then on, things became worse for Hewitt and Walder. Direct appeals for justice to senior staff of the Cancer Research Campaign (CRC) were fruitless. An inspection of the laboratory and its animal houses was proposed by, among others, the Research Defence Society (which campaigns in support of animal-based research and usually concentrates its attention on anti-vivisectionists). The inspection was made by the late Dr Lane-Petter, who told Hewitt personally that the problems seemed to be caused by research staff treating the animal technicians as 'sub-humans'. Hewitt was later informed that the final report was confidential and he would not be allowed to see it; when a copy was leaked to him, he found that blame for the whole situation had been laid, inexplicably, at his door. In the same year, Angela Walder was sacked on 14 counts. Her union took the case to arbitration where it was concluded that she had been unfairly dismissed and should be reinstated; but in the end she was given no practical option except to resign. Hewitt's position became impossible. Colleagues cold-shouldered him, former friends changed their tune; responsibility for the animal house was taken from him, and he was asked to resign as head of the Tumour Radiobiology Section. His health worsening under the strain, Harold Hewitt finally bowed to the pressure on him, and left the laboratory to take up a position at King's College Hospital in London.

Hewitt believes that his criticism of the Director's stance on the welfare of experimental animals, and his support of the technicians who refused to let inhumane practices continue without protest, combined to unleash the fury of the. scientific establishment against him. Unusually, a senior scientist had broken the unwritten code: he had publicly challenged his peers on animal-welfare grounds, and had sided with non-research staff in the process.

A few years later, in the USA, Donald Barnes experienced a situation with some similarities (Barnes 1985). Trained originally as a clinical psychologist, in 1966 Barnes set up a laboratory at the School of Aerospace Medicine in San Antonio, to study the effects of pulsed radiation on the behaviour of (non-human) primates. The monkeys were trained by electric shocks to their feet, of varying and unmeasured intensity. The shocks were increased until the monkeys responded, so that passive animals suffered the highest shocks and death from cardiac fibrillation occurred more than once. Barnes has written that his capacity to use animals in this way was a result of 'conditioned ethical blindness'. Later, Barnes became convinced that the quality of the primate research at the School of Aerospace Medicine was poor, and its relevance to the functioning of aircrews following exposure to nuclear radiation was questionable.

During 1978 he became more vocal in his criticism, and in 1979 he raised objections, mainly on scientific grounds, to an experiment he was ordered to conduct on rhesus monkeys. While his immediate superiors were sympathetic to his scientific criticism, they were frightened to contradict higher authority. That higher authority ordered him to carry out the experiment; he refused, and was sacked. Barnes reacted strongly, convinced that his dismissal was because he had raised animal-welfare issues. He applied successfully for reinstatement, spoke to the press and wrote to humane organisations. In the process, he began to see more clearly the ethical questions involved in animal research, which he had never had the opportunity to consider before. As he wrote, '. . . a bias against any antivivisection philosophy was a "natural" part of the laboratory environment. During my sixteen years in the laboratory the morality and ethics of using laboratory animals were never broached in either formal or informal meetings prior to my raising the issues during the waning days of my tenure as a vivisector.' With sudden insight, Barnes realised the inhumanity of his research and since 1980 he has worked for the humane movement in America.

Scientific conformity is not limited solely to considerations of animal welfare, but is also found in attitudes to methods. For senior scientists, self-interest is a powerful motive for resisting new procedures, since these can undermine the foundations of their seniority and prestige. As Bernard Dixon wrote regarding funding committees who award grants for research (Dixon 1976), such a committee:

> . . . is looking to further the progress of science in a way which will also throw credit on the wisdom with which it has distributed the funds at its disposal. But such cabals are invariably packed with

currently and previously successful practitioners of the branch of science concerned. They are not, therefore, likely to be receptive to heterodox proposals which offend against their own conventional wisdom.

The US National Academy of Sciences report on animal experimentation, commissioned by the National Institutes of Health (NIH), expressed concern that researchers using unconventional systems and models have difficulty finding funding, and recommended that NIH should support good research 'without taxonomic or phylogenetic bias' (National Academy of Sciences 1985).

Many scientists have invested huge amounts of time and energy establishing 'animal models' of human disease, and characterising the physiology and metabolism of various strains of rodent and other species. A new approach to a problem, for example the development of a cell culture technique to study chemical toxicity, requires knowledge at the cell level and the integration of slightly different disciplines—cell biology and biochemistry, electron microscopy, cell culture methods, receptor studies—than are used in whole-animal research. Laboratories have been designed around animals, with investment in appropriate equipment and in training research staff and animal house technicians. The fear of losing status as the 'resident expert' is rarely admitted, but finds expression in other ways, such as in claims that non-animal techniques are insufficiently validated. The fact that much *more* rigorous standards are being applied to humane research than have ever been attained by animal-based techniques is apparently overlooked. A move to *in vitro* or clinically oriented methods requires new apparatus, re-education and new outlooks. This is borne out by an editorial on the subject of animal- and laboratory-based cancer research written by Professor Harrison in a cancer research journal. He said that to question, let alone criticise, accepted principles and 'sacred dogmata' is tantamount to committing public heresy and professional suicide (Harrison 1980).

Dr Frank Perkins, at the time Chief of Biologicals with the World Health Organisation, wrote in 1980 of the inconsistencies and anomalies, tolerated for years by scientists and regulatory authorities alike, in the production and testing of vaccines. Perkins noted that an estimated 120 000 mice, 60 000 suckling mice, 30 000 guinea-pigs and 60 000 rabbits were killed between 1960 and 1980 for double checking on the purity of batches of oral polio vaccine, 'without adding anything to the safety of vaccines' (Perkins 1980). He also highlighted the illogical situation whereby polio vaccine produced from human diploid cell cultures was subjected to the same tests for purity, and *more* control tests, than vaccine which had been produced from monkey kidney cell cultures. Regarding this switch from using monkey cells to human cells for vaccine production, Perkins (1980) wrote:

> There is little point in my reviewing the development of virus vaccines in cell cultures from so called 'primary' monkey kidney cells to the present position of the use of human diploid cell strains. The arguments, personal positions, misconceptions and preconceived ideas that dominated that situation retarded progress for several years. They were arguments that had no more foundations than those put forward against the use of monkey kidney cell cultures on the grounds that every child given the Salk vaccine would have nephritis.

In 1972 the Tuberculosis Reference Laboratory reported (Marks 1972) that growing tissue specimens in culture was a more efficient technique for diagnosing tuberculosis (TB) than injecting tissue samples into guinea-pigs and waiting for infection to develop. The culture method has since been further refined and improved. Nevertheless, as late as 1986 it came to light that the Medical Microbiology Department of the London Hospital was still *routinely* inoculating guinea-pigs for the diagnosis of TB, and in the five years from 1981 to 1985 had used 677 animals for this purpose, the hospital's catchment area including the Borough of Tower Hamlets which has one of the highest notification rates for TB in Britain. Detailed records kept by the hospital allowed a comparison of the results obtained over the period from the guinea-pig and culture tests. Of 677 tests, 616 (91%) were negative by both diagnostic methods; 34 (5%) were positive in both tests; and 22 (3%) were positive by the culture method and negative in the guinea-pig test, confirming Marks' contention that the culture method, when skilfully conducted, is more efficient, perhaps because most environmental mycobacteria are relatively avirulent in guinea-pigs.

In only five cases were results positive in the guinea-pig test and negative by culture technique, and three of these had already been diagnosed by histological examination. Analysis showed that for only one case in the whole series did the guinea-pig test provide data useful to clinicians treating patients. A protracted correspondence with the Home Office was initiated by Clive Hollands, a member of the Animal Procedures Committee, who wanted to know why the London Hospital was still using guinea-pigs for routine TB diagnosis, despite the testimony of several leading microbiologists who agreed that the method is obsolete except in very rare cases, and that the culture technique is more sensitive, cheaper and quicker to perform (Home

Office 1986/87). The Home Office defended the use of the guinea-pig technique in some instances, but at the same time as it concluded in 1987 that it would scrutinise carefully in future all licence applications for this procedure, news emerged that the London Hospital finally decided to stop using guinea-pigs, ostensibly on the grounds of cost.

The development of alternatives to the use of laboratory animals has thus been hindered not only by a lack of funding from orthodox sources, but also by a reluctance on the part of some scientists to be open-minded and apply fair standards to new methods of research and testing.

FREEDOM OF ENQUIRY AND SELF-REGULATION

Intellectual freedom is a cherished concept in the West, and nowhere more so than in science. Scientific and medical progress depend, we are told, on allowing researchers to follow their hunches and pursue lines of investigation which may appear to have no practical outcome. Perhaps it is a fundamental human drive to acquire knowledge about the world and its inhabitants; certainly religious, political, cultural and economic constraints on scientific enquiry have been resisted since the emergence of organised science itself. While it is generally accepted that applied science and technology, because of their direct impact on society, involve numerous ethical decisions, the belief persists that basic research is value-free, and deals only in truths and hard facts, and not in subjective matters or ethical choices. The concept of scientific pursuit being totally objective and free of value judgements is, however, a fallacy. Like other people, scientists must operate within a societal framework. What they choose to study and how they study it are subject to ethical imperatives, whether they are willing to acknowledge these or not. Yet in 1978, Lord Todd, then President of the Royal Society, said 'Ominously, voices have been raised claiming that limits should be set to scientific inquiry—that there are questions which should not be asked and research which should not be undertaken.' Lord Todd was particularly concerned that among the critics had been some scientists (Anon 1978).

Although scientists as a group have always struggled against the imposition of constraints, they have never been entirely free to follow any line of research regardless of other influences, from the mundane restrictions of the availability of funds to the less tangible but more important ethical issues. Few people place freedom of enquiry above good and evil on a scale of priorities, and the scientist who hopes to be left to seek knowledge without the intrusion of moral standards is emulating the ostrich, and an arrogant one at that. When it comes to animal experiments, researchers must face the fact that they too have consciences and can distinguish between good and evil, and are therefore obliged to make moral choices *and* to accept moral imperatives framed by society. If this is true in research in physics and chemistry, how much more so is it essential that ethical issues should be addressed in the biological sciences, where the subjects of research are not sub-atomic particles but living, sentient creatures with a claim to moral concern.

During the deliberations of the Committee of Inquiry into Human Fertilisation and Embryology in 1984, Mary Warnock, who chaired that committee and also the Home Office Advisory Committee on Animal Experiments, drew parallels between the issues each group was exploring. On the subject of human embryo research, she wrote (Warnock 1984):

> . . . matters of morality or ethics are not expert matters, but such as to be solved, if at all by consensus, by an examination of the heart as well as the head . . . But morality is not simply a matter of judging present action against future gain. That is called expediency, not morality.

Between the early years of this century and the end of the 1970s, where were the debates in scientific journals about the ethics of animal research and the conflict of interests inherent in it? Where were the discussions and exchanges which would have indicated that biologists understood the responsibilities they had assumed for the animals they utilised, and were attempting to face these and institute, as a high priority, measures to decrease animal suffering? With a very few honourable exceptions, a conspicuous silence prevailed and the pages of the journals were filled with other, no doubt important, matters. Throughout this period, textbooks on medical ethics embraced an entirely human-oriented approach; research animals and their rights and interests simply did not feature. Only when challenged by the resurgence of a vocal animal-rights movement in the late 1970s, with growing support from the disciplines of philosophy and science itself, and with incontrovertible evidence of animal abuse obtained on video by animal-liberation groups, has the scientific establishment turned its attention to these matters.

On the whole, a scientific education does not provide training in philosophy or ethics, and does not emphasise *connections* between the experience of humans and other animals. As few individual scientists have taken the trouble to educate themselves on the subject of ethics, their grasp of the animal-rights position or indeed the relation of ethics to the use of animals in research has often been feeble.

This is well illustrated in a pamphlet written by Jim Pascoe, a British physiologist active in the defence of animal experimentation, and published by the Physiological Society (Pascoe 1983). Pascoe was seeking to reassure experimental physiologists, especially novices, that their consciences should be their guide to what is acceptable but that the pursuit of good science should be their primary aim. A no-nonsense tone was set from the start: '. . . animal experimentation poses ethical problems, because we are ourselves animals and we have, or possibly imagine we have empathy with other animals' (p. 4). Regarding the ethical dilemma, Pascoe's (1983) views are simplistic and elitist: 'On moral pressures to use less animals, you and your colleagues in the scientific community are the only judges of the numbers needed. It is your reputation; you must stand firm' (p. 11). Finally, his conclusion makes clear the anthropocentric stance from which Pascoe advises young physiologists: 'Ethics I have considered from the angle of protection of the scientist, as one might consider protection from say radiation. Our practice of science and ethics must be firmly based on reason and not be just a blend of hunch and sentimentality' (p. 16).

The issue of self-regulation is an important one. Pascoe has here stated the view, quite commonly held, that experts can be the only rational judges of the need and morality of their actions, and that the notions of lay-people are largely irrelevant because they do not understand the complex issues involved. Serious moral concerns about animal experimentation expressed by lay critics are thus debased, and considered by many researchers as little more than sentimental and emotional outbursts. An ability to empathise with the deprivation and distress undoubtedly suffered by many laboratory animals is disparaged as unscientific and ill-informed. Cora Diamond (1981) wrote of this tactic:

> . . . describing one's opponents as merely expressing their emotions is one way of emphasizing that the issue is not in the moral arena at all, as far as you can see, and the idea that it is a field *for experts* is simply the other side of the same notion . . . if it is urged that the expertise of scientific investigators *is* sufficient and necessary for judging what use of animals in science is appropriate, it is being assumed that there is no significant moral issue.

Michael A. Fox, Professor of Philosophy at Queens University, Ontario, wrote on similar lines in his book, *The Case for Animal Experimentation* (Fox 1986a, pp. 4–5). He claimed that much of the argument for the moral rights of animals has been characterised by sentimentalism of two sorts—the traditional 'heart-strings' variety and the 'modern intellectualised' type: 'The latter is frequently present in the work of speculative ecologists, theologians, and academics who have joined the animal rights movement but appear to possess scanty firsthand knowledge of animals or the uses to which they are put by humans'. Here, using the charge of ill-informed sentimentality, Fox was desperately trying to dismiss the concerns of a new breed of more serious opponents—the articulate and academic advocates of animal rights, who were forcing the question of the status of experimental animals into the moral arena.

Interestingly, within months of the publication of his book, Fox publicly announced his retraction of its basic premise, that animal experimention is justified by the needs and concerns of humans. In a letter to *The Scientist*, an American science newspaper, Fox agreed with a reviewer of his book that its philosophical approach was superficial, dogmatic and unconvincing. He had, he wrote, come to be 'profoundly dissatisfied' with the narrow definition of the moral community used in his book, and now believed that efforts to justify the use of animals for experimentation would convince no-one except the already converted (Fox 1986b). Fox has since written in some depth on why he changed his views about animal experiments. In answering his own question as to how someone of his intelligence and sensitivity could have held the views expressed in his book, Fox cites factors as applicable to a scientist as to a philosopher: personal advantage, social conditioning and the approach used in teaching ethics, which encourages objectivity, disinterestedness and impartiality to the exclusion of feeling and emotion (Fox 1987).

It is not sensible, as Stephen Clark pointed out (Clark 1977, pp. 141-2), to expect that human decency will always prevail in the face of peer-group pressure, desensitisation to suffering, job security, the powerful drive of scientific curiosity and the lure of scientific eminence. In general, society imposes tight restrictions on the investigations which may be carried out on humans, because the rights of the individual are higher on the scale of priorities than scientific or even medical gain. The permission of a local ethical committee must always be obtained in the UK before human volunteers are used in any research, no matter how harmless it may appear; how much more, then, is there a need for ethical judgements to be made to protect laboratory animals on whom it is sometimes expedient to perform severe and painful procedures, who are not volunteers, cannot speak for themselves and are not protected by being of the same species as us?

During World War II, the pursuit of knowledge led dozens of German and Japanese scientists and doctors to forget personal responsibility for good and evil and to vivisect Jews in Nazi Germany and prisoners of war in Manchuria. Simple humanity, moral standards, empathy and sentimentality were swept away, and self-regulation failed utterly. Even though this research enabled Japan, for example, to lead the field in the production of penicillin, in understanding the role of vitamins, human adaptation to low temperatures and the development of artificial blood (Whymant 1983), it is universally condemned. It is inadequate to claim that those scientists were entirely evil, corrupt people, quite different from researchers of today, who labour for the good of humanity and whose work unfortunately requires animals. The ethics and practice of animal experimentation are matters for all concerned people, not just the 'experts'.

The failure of self-regulation in the scientific community with respect to animal experiments can be seen most clearly in the USA. There, animal-liberation groups have been particularly successful in revealing animal abuses, for example Edward Taub's deafferentation studies at the Silver Springs Institute of Behavioural Research, where monkeys were denied basic veterinary care, and the University of Pennsylvania head injury research, where inadequately anaesthetised primates were operated on in unsanitary conditions, and treated with contempt. Reports by the US Department of Agriculture inspectors make horrendous reading, describing rusty cages with sharp protrusions, often being too small to allow animals to make postural adjustments, insufficiently regular inspections, build-up of faeces in cages and contamination of feed by urine. These reports deal with numerous research institutes at places such as Harvard, Yale and Johns Hopkins, which were approved by the American Association for the Accreditation of Laboratory Animal Care and funded by the National Institutes of Health to the tune of several million dollars a year (Stevens 1984). Two major differences between America and the UK are that there is stricter legislative control of animal experiments in the UK which, however, does not have a Freedom of Information Act—so reports of inspectors are not available to the public. Unless British scientists believe that they are inherently more sensitive to animal welfare than their American colleagues, the different situations in the two countries indicate that it is detailed and enforced legislation on standards of animal care and use which is the major factor in ensuring humane treatment of laboratory animals, rather than self-regulation.

One result of animal-rights campaigns in the 1980s has been the long-overdue publication in scientific journals of several articles tackling the subject of animal experiments, as impending legislation in the USA and the UK sharpened scientists' interest. Some of the articles have been little more than rallying cries to the professions to unite in combat against animal-rights and animal-welfare pressures, with scant interest in a fair debate (see, for example, Rosner 1985, Berliner and Barger 1986, Smith and Hendee 1988). In particular, Grubb defended the now-ended head injury research on primates at the University of Pennsylvania, in the face of almost universal condemnation from scientists, politicians and the public on both sides of the Atlantic, and complained that the abuse of animals revealed by animal-liberation raids had resulted in surprise inspections by NIH and occasional withholding of funds (Grubb 1987). Others have attempted to tackle the issues with more open minds (see, for example, Hoff 1980, Britt 1984, Iglehart 1985, Overcast and Sales 1985).

Another welcome, although belated, response from the scientific community has been the publication of guidelines for researchers, especially those intending to submit papers to journals. Researchers are expected to adhere to the guidelines, and their papers may be rejected if they have not. Guidelines have been published in the journal *Animal Behaviour* (ASAB and Animal Behaviour Society 1981) and similar advice was given to experimental psychologists in two documents from the British Psychological Society (1979, 1985). *Pain* (Zimmerman 1983) and other journals have done likewise, sometimes acknowledging the ethical dilemmas posed by animal experimentation (*ICRS Journal of Medical Science* 1981). A further indication that the issue is *beginning* to receive attention is that, since 1985, *Index Medicus* has included 'Animal testing alternatives' in its subject headings, and, since 1986, 'Animal welfare' has also figured, with a list of some 76 publications appearing in the cumulated *Index* for that year.

CLAIMS AND COUNTER-CLAIMS

Scientific organisations and individuals who defend animal experimentation have criticised the animal-welfare and anti-vivisection movements for conducting campaigns with misleading information, for oversimplifying the 'facts' and for other unprofessional tactics. Yet pro-vivisection propaganda has frequently contained distorted rhetoric and bias as well as misinformation. On the question of pain and the use of anaesthetics, the anti-vivisection movement has been criticised, rightly on occasion although less so now, for giving the impression that all experiments cause severe pain and that surgery is performed on unanaesthetised animals. However, pro-vivisection groups and individuals have been guilty of attempting to mislead on this crucial subject in efforts to alleviate public concern.

Any British scientist who conducts animal experiments should know that, under the recently superseded Cruelty to Animals Act 1876, anaesthetics did *not* need to be given (except in the case of surgery) if they would frustrate the object of the experiment—even if pain resulted. More than 80% of all animal experiments in Britain did not involve the use of anaesthetics, including most toxicity tests such as the LD50 test, which may cause severe pain, according to the Home Office Advisory Committee in their *Report on the LD50 Test* (Home Office Advisory Committee 1979). Under the Animals (Scientific Procedures) Act 1986 a similar system continues, in that a researcher must prevent or reduce pain, distress and discomfort to a minimum when this is 'consistent with the purposes of the authorised procedures' (Section 10 (1) (2) (a) of that Act). Yet a leaflet produced and circulated by the drug-company-funded Animals in Medicines Research Information Centre (AMRIC), with a supporting message from the Chair of AMRIC's Advisory Board, Professor Sir John Butterfield, claimed that 'more than 80% of experiments do not require an anaesthetic *because they are not likely to be painful*' [my emphasis] (AMRIC 1984).

In another AMRIC leaflet it is stated that 'Anaesthetics and analgesics are used when necessary' (AMRIC undated), implying that animals are never allowed to suffer pain. In the case of analgesics, in particular, this statement is laughable, as pain-killers are virtually never given to rodents and not always to animals such as cats, dogs and primates. In an article in a newsletter of the pro-vivisection Research Defence Society (RDS), abridged from one written by Professor Tim Biscoe (an experimental physiologist and honorary secretary of the RDS) for an Oxford University current affairs magazine, Biscoe wrote: 'Experiments on whole animals whose systems are studied will involve pain no more than is necessary to give an injection of anaesthetic from which the animal will not recover' (Biscoe 1985). While this is true for some non-recovery experiments in physiology research, this particular article was addressing the general use of animals in medical research, and in the absence of clarification thereby gives the impression that no animals suffer pain in medical research, which is very far from the truth.

In the literature defending animal experiments, as already discussed regarding research reports in the journals, grammatical conveniences abound, with the effect of distancing the reader from the reality of animals suffering in research. Animals are ascribed willing intent to help humanity by somehow volunteering themselves for experimentation, as in the RDS leaflet which describes advances in the treatment of diabetes 'and the role *animals have played* in research' [my emphasis] (Research Defence Society 1982a). Similarly in a piece of propaganda from AMRIC we can read that, until alternative techniques are available for drug development, 'animals *must help prove* their safety and efficacy' [my emphasis] (AMRIC 1984). AMRIC has also produced a booklet for children under 13 years old, called *Imagine What Life Would Be Like Without Animals* (AMRIC 1987). This explains that, in the case of people with incurable diseases, 'We can only learn how to help them *with the help of animals*' [my emphasis].

One well-used defence of animal experiments is to intimate that there is a medical relevance to virtually any piece of research. This approach is heavily exploited in applications made to granting bodies for funds, but is also used in letters, articles and leaflets in general defence of animal experimentation. Another ploy is to suggest that animals themselves may be major beneficiaries of research, even when it is directed towards human problems. This argument, a fairly unconvincing justification, was used by Professor Patricia Scott in a debate at the London School of Economics in 1981: 'Even where experiments are specifically designed to be advantageous to human beings, there is nearly always some degree of "spin-off" which leads to improvements in the care and treatments of accidents and diseases in members of the species being used for investigation' (Scott 1981). Since 80% of animals used in research are rodents, as we are so often told by the RDS, it is astonishing that Professor Scott believed that there is so much concern for treating injuries and disease in those species. Henry Salt revealed the fallacy some 90 years ago, expressing doubts that even if animals could understand the logic, they would hardly rush 'eagerly onto the knife'. Applying the same argument to humans, he noted that if the case for broad benefits to others than the subjects of research was so meritorious, why were there so few human volunteers coming forward 'to die under the hands of the vivisector?' (Salt 1892). As Professor Stephen Clark points out in his book (Clark 1977, pp. 127-8), would we have approved blithely the Nazi experiments on Jews on the grounds that some Jewish people, somewhere, might have benefited?

Anti-vivisectionists have been accused of misanthropy and of caring only about animals, ever since organised protests against animal experiments first began. An example of the theme occurred in Michael A. Fox's book: 'Lamentably, it often seems that among the most vociferous proponents of animal welfare there is an abundance of concern for the plight of other species and very little concern for that of our own' (Fox 1986a, pp. 4–5). Similarly, Dr Brian Meldrum of the Institute of Psychiatry is reported as telling a delegation from the Christian Consultative Council for the Welfare of Animals (CCCWA) that they had their priorities wrong in worrying about the inadequate caging of baboons in his laboratory, when there were so many homeless people needing help (CCCWA 1987). These are unjustified criticisms, as even a superficial study of the history of anti-vivisection will show. Frances Power Cobbe, who founded the anti-vivisection movement in Britain in the 19th century and led it with great vigour, was a social worker and advocate for university education for women. Nearly 100 years ago, Henry Salt, humanitarian and writer, campaigned with passion against flogging in the Navy, against capital and corporal punishment, and for reform of prison conditions. He was as well known for his opposition to all forms of animal exploitation, from vivisection to the fur and feather trades, and spoke with clarity in favour of animal rights; and the Earl of Shaftesbury's social reforms are better remembered than his no less vigorous opposition to animal experiments.

Dr Walter Hadwen, a leading anti-vivisectionist at the turn of the century, was a general practitioner who let patients convalesce, sometimes at his own expense, in his country home; and who was instrumental in essential improvements in sanitation, water supply and transport systems in Highbridge, Somerset, where he lived for many years. The same connection between concern for animal and human welfare and rights is evident in key characters today: Lord Houghton's championing of animals in Parliament has been complemented by his work as one-time Minister for the Social Services, as Chair of an earlier Teachers' Pay Review Body and by his long-standing interest in family planning issues. Peter Singer, whose book *Animal Liberation* (Singer 1976) helped trigger the revival of the animal protection movement in Britain, has been moved by a sense of universal justice to write of ethical issues relating to humans as well as animals (see, for example, Singer 1979). Many members of animal-rights groups also campaign actively *against* racism and torture, and *for* peace and human rights; they are aware that justice and compassion are indivisible.

The discerning public's appreciation of scientific logic must have taken a knock from repeated statements attempting to justify animal experiments by drawing attention to the numbers of unwanted cats and dogs put down by the Royal Society for the Prevention of Cruelty to Animals each year, as if the one excused the other (Huxley 1983, p.iii, AMRIC 1984). Efforts by the RDS to defend highly criticised cosmetics tests on animals caught them up in such logical nonsense as: 'Humans differ so widely that very large numbers would be necessary to obtain reliable results. However, specially-bred laboratory animals do resemble each other to a very great degree and reliable tests can be made on fewer laboratory animals than human beings' (Research Defence Society 1982b). Those whose minds are not straitjacketed by scientism will see immediately that using inbred, genetically similar animals may produce consistent results, but is unlikely to improve the relevance of tests to outbred, genetically dissimilar humans!

The anti-vivisection movement is accused of choosing emotionally charged photographs and words, while scientists supposedly restrict themselves to objective statements of fact. Two points arise here, and the first is the assumption that bringing emotions into the subject of animal experimentation is somehow unreasonable. As discussed earlier in this essay, while objectivity has its place, an ability to empathise with others, even of another species, is important and not reprehensible. No-one disputes the use of pictures of starving children by charities concerned with the Developing World, or descriptions of their plight in language which is other than coldly scientific. Human and animal welfare are issues which demand the exercise of our emotions. The use, by several medical charities, of photographs of handicapped or sick children shows that they realise this. The pictorial emphasis on infants and children in advertisements by the Arthritis and Rheumatism Council for Research, even though these diseases primarily afflict the middle-aged and elderly, can be compared with the use by anti-vivisection societies of photographs of cats and dogs under experimentation, even though rodents are the most common laboratory animals. Neither is strictly incorrect, but if scientists criticise one then consistency demands that they should also speak out against the other.

SUMMARY AND CONCLUSIONS

I have attempted to show that the way science is taught, its hierarchical structures and the conformity of its practitioners, has prevented advances in laboratory animal welfare which could otherwise have been implemented much sooner. The insular and reactive nature of the scientific community, on this subject in particular, has also suppressed free debate on the morality of using animals as tools in research, testing and education—a debate which has now been forced on scientists by an increasingly professional and vocal anti-vivisection movement using a philosophy of animal rights. Researchers are beginning to emerge from their complacency and tackle questions of morality, but public confidence in the scientific community has been shaken. While self-regulation is to be encouraged, well-enforced legislation with a means to ensure public accountability is essential.

Equally important is the need to introduce discussions of animal rights and animal welfare to students taking science, philosophy and humanities courses at colleges and universities, and to encourage a greater awareness of the social and moral context of scientific research. Unjust and inaccurate criticisms of the animal protection movement by animal research defence organisations do not engender respect for the scientific community or reflect well on its members; neither do inflated claims for the medical benefits of animal research. A determined effort by scientists to 'put their house in order ' might yet take some of the bitterness out of the conflict over animal experimentation, and allow further progress to be made in the reduction, replacement and refinement of animal experiments and in a sensitive appreciation of laboratory animals as individuals whose interests demand consideration. As civilisations advance, the trend is always towards increasing humanity and compassion, and decreasing cruelty and exploitation. The instinctive kinship which we feel for other animals and the natural world is being encouraged, and injustices which failed to move earlier generations have been ended. Public conscience about animal experiments has been well and truly awakened; scientists must respond to this, and encourage reform from within the scientific community.

REFERENCES

AMRIC (1984). *Animals in Medicines Research—Why?* (leaflet), AMRIC, London

AMRIC (1987). *Imagine What Life Would Be Like Without Animals* (booklet for schools), AMRIC, London

AMRIC (undated). *Animals in Medicines Research—Now Read the Facts* (leaflet), AMRIC, London

Anon (1978). Royal Society President questions anti-science dogma, *New Sci.*, **80**, 748

ASAB and Animal Behavior Society (1981). Guidelines for the use of animals in research, *Anim. Behav.*, **29**, 1–2

ASE/IOB/UFAW (1986). The place of animals in education, *Biologist*, **33**, 275–8

Barnes, D. J. (1985). A matter of change. In Singer. P. (ed.), *In Defence of Animals*, Basil Blackwell, Oxford, pp. 157–67

Berliner, R. W. and Barger, A. C. (1986). Physiology and antivivisection, *News Physiol. Sci.*, **1**, 79–80

Biscoe, T. J. (1985). The use of animals in medical research, *RDS Newsl.* May, 2–3

British Psychological Society—Scientific Affairs Board (1979). Report of the Working Party on Animal Experimentation, *Bull. Br. Psychol. Soc.*, **32**, 44–52

British Psychological Society—Scientific Affairs Board (1985). Guidelines to the use of animals in research, *Bull. Br. Psychol. Soc.*, **38**, 289–91

Britt, D. (1984). Ethics, ethical committees and animal experimentation, *Nature*, **311**, 503–6

CCCWA (1987). *Report on a Visit to the Institute of Psychiatry*, CCCWA, London

Clark, S. R. L. (1977). *The Moral Status of Animals*, Clarendon, Oxford

Delacôte, G. (1987). Science and scientists: public perception and attitudes. In Evered, D. and O'Connor, M. (eds), *Communicating Science to the Public*, John Wiley, Chichester, pp. 41–8

Diamond, C. (1981). Experimenting on animals: a problem in ethics. In Sperlinger, D. (ed). *Animals in Research: New Perspectives in Animal Experimentation*, John Wiley, Chichester, pp. 337–62

Dixon, B. (1976). *What is Science For?*, Penguin, London, p. 134

Fox, M. A. (1986a). *The Case for Animal Experimentation: An Evolutionary and Ethical Perspective*, University of California Press, Berkeley, CA

Fox, M. A. (1986b). Author reverses views on animal rights (letter), *The Scientist*, 15 December, 10

Fox, M. A. (1987). Arguing against animal research, *The Whig Standard*, Canada, 4 April, 5–7

Goodall, J. (1971). *In The Shadow of Man*, Houghton Mifflin, Boston, MA

Goodall, J. (1987). Prisoners of science: chimpanzees in medical research, *Bull. Psychol. Ethical Treatm. Anim.*, Spring, 2–5

Grigsby, P. and Maruyama, Y. (1981). Modification of the oral radiation death syndrome with combined WR-2721 and misonidazole, *Br. J. Radiol.*, **54**, 969–72

Grubb, R. L. (1987). Animal rights versus medical research, *Neurosurg.*, **20**, 809–10

Harrison, D. F. N. (1980). Editorial, *Clin. Oncol.*, **6**, 1–2

Heim, A. (1981). The desensitization of teachers and students. In Paterson, D. A. (ed). *Humane Education—A Symposium*, HEC, Sussex, pp. 37–45

Hoff, C. (1980). Immoral and moral uses of animals, *New Engl. J. Med.*, **302**, 115–18

Holden, C. (1987). Industry toxicologists keen on reducing animal use, *Science*, **236**, 252

Home Office (1986/87). Personal correspondence with Clive Hollands

Home Office Advisory Committee (1979). *Report on the LD50 Test*, Home Office, London, p. 13

Huxley, A. (1983). Anniversary address by the President (of the Royal Society), *R. Soc. News Suppl.* **2**, i–vii

ICRS *Journal of Medical Science* (1981). Six principles of humane animal experimentation—Editorial, *ICRS J. Med. Sci.*, **9**, 277–9

Iglehart, J. K. (1985). The use of animals in research, *New Engl. J. Med.*, **313**, 395–400

Kelly, J. A. (1986). Alternatives to aversive procedures with animals in the psychology teaching setting. In Fox, M. W. and Mickley, L. D. (eds), *Advances in Animal Welfare Science 1985/86*, HSUS, Washington, DC, p. 168

Langley, G. R. (1981). Animals in British universities. In Paterson, D. A. (ed.), *Humane Education—A Symposium*, HEC, Sussex, pp. 25–35

Marks, J. (1972). Ending the routine guinea pig test, *Tubercle*, **53**, 31–4

Medawar, P. B. (1969). *The Art of the Soluble*, Pelican, London

Morley, D. (1978). *The Sensitive Scientist: Report of a British Association Study Group*, SCM, London, pp. 84–91

National Academy of Sciences (1985). *Models for Biomedical Research*, National Academy Press, Washington, DC

Overcast, T. D. and Sales, B. D. (1985). Regulation of animal experimentation, *J. Am. Med. Assoc.*, **254**, 1944–9

Pascoe, J. E. (1983). *Attitudes to Experimentation on Living Animals: Science, Ethics, Law*, Physiological Society Education Sub-Committee

Paton, W. (1984). *Man and Mouse: Animals in Medical Research*, Oxford University Press, Oxford

Perkins, F. T. (1980). Risks and gains associated with vaccination, *Dev. Biol. Standard.*, **46**, 3–13

Research Defence Society (1982a). *Diabetes: Research Triumphs* (leaflet), RDS, London

Research Defence Society (1982b). *Are Animal Experiments Necessary for Cosmetic Products?* (leaflet), RDS, London

Rollin, B. E. (1981). *Animal Rights and Human Morality*, Prometheus Books, New York

Rosner, F. (1985). Is animal experimentation being threatened by animal rights groups?, *J. Am. Med. Assoc.*, **254**, 1942–3

Rowan, A. N. (1984). *Of Mice, Models, and Men: A Critical Evaluation of Animal Research*, State University of New York Press, New York, p. 16

Russell, W. M. S. and Burch, R. L. (1959). *Principles of Humane Experimental Technique*, Methuen, London

Salt, H. S. (1892). *Animals' Rights Considered in Relation to Social Progress*; reprinted (1980), Society for Animal Rights, Clarks Summit, pp. 98–9

Scott, P. (1981). Speech in a debate at the London School of Economics, 11 March

Shapiro, K. J. (1987). A student's right to a careful education, *Bull. Psychol. Ethical Treatm. Anim.*, Fall, 9–11

Shuster, S. (1977). Why we need animal research, *World Med.*, **13**, 19–37

Singer, P. (1976). *Animal Liberation*, Jonathan Cape, London; reprinted (1983) Thorsons, Wellingborough

Singer, P. (1979). *Practical Ethics*, Cambridge University Press, Cambridge

Smith, S. J. and Hendee, W. R. (1988). Animals in research, *J. Am. Med. Assoc.*, **259**, 2007–8

Stevens, C. (1984). Mistreatment of lab animals endangers biomedical research, *Nature*, **311**, 295–7

Warnock, M. (1984). Scientific research must have a moral basis, *New Sci.*, **104**, 36

Whymant, R. (1983). The butchers of Harbin, *Conn. Med.*, **47**, 163–5

Zimmerman, M. (1983). Ethical guidelines for investigations of experimental pain in conscious animals, *Pain*, **16**, 109–10

10. Legislation and the Changing Consensus

Judith Hampson

'There is a notable reluctance among (British) scientists to discuss the moral implications of animal experimentation, possibly because, by and large, they are uncomfortable in the area of philosophical (especially ethical) debate. As long as their work is within the law, what research scientists do is very much their own responsibility and perhaps ethical issues are a matter for individual conscience, rather than open debate. The growing interest and concern of a broad section of the rational general public renders this position untenable today.'

David Britt (1984). Nature, 311, 503

INTRODUCTION

Laws designed to control the practice of experimentation on living animals have a central purpose: 'to reconcile the needs of science with the just claims of humanity'. This purpose was outlined by the UK Royal Commission on the Practice of Subjecting Live Animals to Experiments, the first enquiry of its kind, which led to the earliest law in the world controlling the practice, the Cruelty to Animals Act 1876. Since that time, legislation to control animal experiments has been enacted in many countries throughout the Western world. But a century of heated debate on this thorny topic has resulted only in changes to the details of such laws and to their administration, not to their fundamental purpose and scope. This remains everywhere the same as that of the UK 1876 Act—to restrict experimentation within what society deems to be acceptable limits while causing least harm to free scientific enquiry.

These laws rest on a basic assumption, that it is legitimate and morally acceptable to kill animals and to cause them pain and distress in order to protect human society from illness and environmental hazard and to gain scientific knowledge. Most laws seek to impose limits upon the degree of pain and distress which might be caused and some seek to restrict the purposes for which it is considered legitimate, but none question its legitimacy. The only national law which might be considered an exception to this rule is the

219

1936 Animal Welfare Act of Liechtenstein, article 3 of which prohibits vivisection, though a new draft will allow animal experiments to be conducted by government agencies in exceptional circumstances.

The assertion that the use of animals in research is morally reprehensible as a means and can never be justified, whatever the ends, has always lain at the heart of the anti-vivisection movement. Society as a whole has remained unmoved by this claim and it is only in the last decade that a rigorous intellectual challenge has been mounted against the *status quo*. It has been issued by a small number of philosophers around the world who have sought to establish the immorality of using animals, in all or most instances, as research tools (Singer 1975, 1985, Regan 1983, Rollin 1981, Clark 1977, Midgley 1983). Whether society will respond to this challenge by establishing a new moral agenda whereby animals are acknowledged as legitimate objects of moral concern, and accorded concomitant legal rights, remains to be seen. This is a question for society as a whole to answer. This essay will look at some of the legislation which has recently been enacted to control this usage of animals and will discuss the extent to which it might be expected to achieve some of its major objectives.

CURRENT LEGISLATIVE INITIATIVES

So long as society continues to endorse animal experiments, the laws which aim to protect research animals will be designed to keep their use within defined limits and controlled under specified conditions. There are three elements which are central to any effective system of control, although this essay will argue that no system currently in place has successfully implemented all of them. These are: a legislative framework; an ethical and scientific review system; and specified rules for the care and use of animals. All regulatory systems have five main aims:

● To define legitimate purposes for which laboratory animals may be used.
● To exert control over allowable levels of pain or other distress.
● To provide for inspection of facilities and procedures.
● To ensure humane standards of animal husbandry and care.
● To ensure public accountability.

Many Western countries are currently passing new laws, updating old ones or struggling to improve their systems of administration in ways which attempt to fulfil at least some of these aims. Pressure has been placed on governments to do this largely by public opinion. In the last decade the topic has seldom been out of the media. In the

United Kingdom a concerted campaign by leading animal-welfare and animal-rights groups to 'put animals into politics', spearheaded by Lord Houghton, took place in the 1977 run-up to a general election and culminated in the repeal of the Cruelty to Animals Act 1876 and the passage of the Animals (Scientific Procedures) Act 1986. Similar activity was taking place in a number of European countries, notably the Netherlands, the Federal Republic of Germany, Belgium, Switzerland and the Scandinavian countries, all of which have passed or amended legislation on animal experimentation in the last decade.

One impetus to this activity has been the Council of Europe Convention and the subsequent Directive drawn up by the Commission of the European Economic Community, which have provided frameworks within which individual member countries should operate. Another has been the set of international guiding principles for bio-medical research throughout the world, drawn up by the Council for International Organisations of Medical Sciences in Geneva (CIOMS 1985).

In the USA recent changes have been equally extensive and the debate over new controls has been set against an even more colourful backdrop of animal-rights activity than in European countries. Over the last few years exposés by animal-rights activists, often the result of meticulous underground investigations and illegal break-ins, have highlighted glaring inadequacies in the US system of control.

These events have played a central role in prompting recent initiatives by Congress and by major research funding agencies in the USA to implement extensive legislative and administrative controls.

Legislative frameworks

Though there are many laws covering animal experimentation throughout the Western world, which must be tailored to the needs of individual countries, there are two major types of legislative framework. The one with the longest history is the type operated in the United Kingdom, which is based on a strictly defined control system, involving licensing and inspection, operated by a government department which defines guidelines for conduct under the legislation and oversees compliance of institutions and individuals.

The other type, currently being developed in countries covering large geographical areas with widely dispersed research facilities, such as the USA, lays down a basic legislative framework within which the controls are provided and implemented at the institutional level through Animal Care and Use Committees. Canada has a long-established regulatory system based on institutional committees (Rowsell 1987) and Australia is currently developing a similar system.

CONTROL BASED ON NATIONAL LICENSING AND INSPECTION: THE UK ANIMALS (SCIENTIFIC PROCEDURES) ACT 1986

The UK, with its exceptionally long history of control, has perhaps the most elaborate and tightly regulated system in the world. The new Act, which passed onto the Statute Books in June 1986, is a piece of enabling legislation specifically prohibiting very little. Rather it aims to establish stricter controls through its administrative machinery, which is explained in detail in a document published by the Home Office—*Home Office Guidance on the Operation of the Animals (Scientific Procedures) Act 1986* (Home Office 1986). The mechanism by which the Act operates cannot be understood simply by reading the Act itself, it must be read in conjunction with the Home Office document.

Regulated procedures, protected animals and licensing

Under the 1986 Act provisions are extended to cover not only 'experiments' but also non-experimental techniques such as the production of vaccines and sera, passage of tumours and breeding for physical defects liable to cause suffering, as well as techniques such as decerebration, none of which were covered by the 1876 Act.

The definition of a 'protected animal' has also been extended to include not only all vertebrates but also foetal, larval and embryonic forms from the halfway point of gestation or incubation in the case of mammals, reptiles and birds and from the point when they are capable of independent feeding in respect of fish and amphibians. All regulated procedures on protected animals must be conducted under the authority of two licences, a personal licence and a project licence.

New personal licences require the testimony of a sponsor who will usually be a senior licensee in a position of authority in the applicant's institution and is able to testify that the applicant has had an appropriate education and training, knows the relevant techniques for the species to be used, understands the necessary requirements for care of the animals and, where appropriate, has demonstrated suitable proficiency in laboratory animal practice. Personal licences will be issued after the local Inspector has interviewed the applicant and assessed his or her competence, particularly in anaesthesia, analgesia, euthanasia and animal care. The personal licence does not give authority to carry out a procedure but merely defines the types of procedure and animals for which the licensee is covered under the authority of a project licence.

The project licence is a long and complex document requiring detailed description of the procedures to be carried out and the justification for the work. It may, on occasions, cover only one particular piece of experimental work and may be held by the same person holding the personal licence. In many instances, however, it will cover large programmes of work, such as the range of toxicity tests carried out in an industrial company on a particular type of product. The project licence holder is responsible for the scientific direction and overall administration of the project, on which a large number of personal licensees (which may or may not include the project licence holder personally) will be working. Where necessary, an assistant project director will also be named.

With project licence applications the Inspector may recommend that the Home Secretary seek also the opinion of an independent assessor, drawn from a list available to the Home Office of technical advisers from all branches of the biological sciences. Referral to such an assessor, to whom the applicant will first be asked if there is any objection, will take place if the techniques to be covered are highly specialised or novel, or if for any other reason the proposed project gives the Inspector cause for concern.

Pain and purpose

The new licensing system gives the Home Office a much greater degree of control over the amount of pain or suffering permitted, and the purposes for which it is allowed. Each project is given a severity banding—mild, moderate or substantial. The banding refers to the overall project rather than to the experience of individual animals, but the number of animals expected to suffer considerable ill-effects will obviously affect the overall category. Home Office Inspectors are working together on the grading of projects to ensure a level of consistency.

Purposes for which a project licence can be granted are drawn very widely, in line with those outlined by the European Convention, so as to bring under the Act as many types of procedure as possible. However, crucially, the Act provides for a cost–benefit analysis linking the permissible amount of pain or distress to the purpose of the project (Animals (Scientific Procedures) Act 1986, Section 5 [4]):

> In determining whether and on what terms to grant a project licence the Secretary of State shall weigh the likely adverse effects on the animals concerned against the benefit likely to accrue as a result of the programme to be specified in the licence.

In the same section, the Act also requires that the applicant has given adequate consideration to the feasibility of achieving the purpose of the programme without the use of protected animals before applying for the project licence.

The infliction of pain is further controlled by conditions which may be attached to the personal licence as deemed necessary by the Secretary of State. Two particular conditions will invariably be attached, one requiring the minimisation of pain, distress and discomfort, and an inviolable condition specifying occasions when the procedure must be terminated and the animals humanely killed.

The actual wording of this termination condition is given in the Home Office guidance notes (Home Office 1986):

> In all circumstances where an animal which is being or has been subjected to a regulated procedure is in severe pain or severe distress which cannot be alleviated the licensee must ensure that the animal is painlessly killed forthwith by a method appropriate to the animal specified in Schedule 1 of the Act or by such other method as may be authorised by his [*sic*] licence.

Not only is the new termination condition stricter in that it requires immediate killing of any animal suffering severe pain or distress, whether or not the object of the experiment has been attained or the suffering is thought likely to endure, but the condition is actually specified on the face of the statute. This is a provision for which the animal-welfare movement fought long and hard during the passage of the Bill. Though the term 'severe' cannot be simply defined in the Act itself, this specification of the condition in the statute, together with the severity banding system for project licences, makes it necessary for some consensus to be reached between Inspectors and experimenters regarding the definition of these terms and what kinds of procedure are described by them. Such a consensus can only be reached by detailed consideration of specific examples (see the Discussion section of this essay).

These provisions provide a mechanism whereby the Home Office can scale down the number of experiments involving severe procedures, and the Minister gave an assurance during the Bill's passage that (**Hansard** 1986):

> only a limited number of procedures will be allowed to approach the old   pain
> condition and others will be scaled down to lower categories.

Inspection and standards of care

All experimental and breeding facilities holding laboratory animals must be registered under the Act and must comply with the conditions specified on their registration certificate. They are open to periodic inspections by the Home Office Inspectorate. Such visits are usually unannounced, though the Inspectorate works by forming a co-operative relationship with laboratory personnel rather than operating as a police force, a system which has sometimes been criticised. Crucial to the implementation of this new system of control are the provisions for ensuring daily care of the animals in the laboratory itself. The registration certificate will name three persons specifically. One is a person in a position of authority who is ultimately responsible for ensuring compliance with the conditions of the certificate; the second is the person responsible for day-to-day care of the animals, usually a highly trained chief animal technician or animal house curator; the third is a named veterinary surgeon, usually employed by the institution, or in the case of small establishments, contracted to it.

These latter two persons have responsibility for informing the personal licensee if an animal in his or her charge is giving cause for concern and, if the licensee is unavailable, they have the authority to take steps to see that such an animal is adequately cared for or, if necessary, killed.

Further new measures contained in the 1986 Act include a clause permitting the re-use of animals which have already undergone a general anaesthetic, but only when certified by a veterinary surgeon that they are in a good state of health, will not be kept for an excessively long time between the first and second procedures, and the second procedure will be under terminal anaesthesia. Even such restricted re-use requires the specific authority of the Secretary of State. Another tightening up has been the bringing of breeding and supplying establishments under the system of control and inspection with the added provision, pressed during the Bill's passage by the RSPCA (Royal Society for the Prevention of Cruelty to Animals), that (Animals (Scientific Procedures) Act 1986, Section 10 [3,a]):

no cat or dog shall be used under the licence unless it has been bred at and obtained from a designated breeding establishment.

Exemptions are to be made only where no suitable animal is available from registered breeders; for example, for research into diseases affecting specific breeds of dogs. This new stringent provision, along with the systems of marking and record keeping also required, will dispel, once and for all, public apprehension that stolen and stray animals have found their way into British laboratories in the past.

Accountability

A further measure of accountability is provided in the system by means of the new Animal Procedures Committee (APC) which, unlike its predecessor, the Home Office Advisory Committee, under the 1876 Act, has statutory authority. The Committee currently consists of a Chairman who is a Professor of Law and 19 members. The Act provides that no more than half the members should be licensees, or have held a licence in the past six years, and that at least two-thirds of its membership will be provided by medical practitioners, veterinary surgeons or biologists. Animal-welfare interests must also be adequately represented. Currently the Committee has among its membership three professional animal welfarists, as well as a QC who is Chairman of the legal committee of the RSPCA and of the well-known Battersea Dogs Home, and a Canon who has specialised in the study of ethics.

The Act invests the Committee with the power to investigate and report on such matters as *it* may determine, as well as upon those requested by the Secretary of State. It will have referred to it any licence applications which the Home Secretary, usually on the advice of the Inspectorate, regards as unusually controversial or which present peculiar technical difficulties.

In addition to these it also will have referred to it applications giving rise to particular public concern, including applications for the testing of cosmetics and toiletries and all research on tobacco and its substitutes (unless carried out under terminal anaesthesia), both areas that are the subject of long-standing public debate.

The Committee has also been asked to give an opinion on the granting of all licences for the purpose of gaining manual skill in microsurgery. This is an area of public concern because it represents a widening of permissible purposes compared with the 1876 Act, which specifically excluded procedures carried out for the purpose of gaining manual dexterity. Permission is being granted only in respect of work carried out on terminally anaesthetised rodents, and all projects require the approval of the APC.

The Committee will also see, retrospectively, all applications submitted in the category of substantial severity, and it will thus be able to make recommendations for restriction or modification of similar projects in the future. Another part of the Committee's brief is to review general trends in research, advise on the mechanism of the control system and investigate areas of special public concern. Currently it is investigating behavioural and psychological research, a function handed down by its predecessor, the Home Office Advisory Committee.

Public accountability of the workings of the Committee is ensured since the Act requires its Annual Report to be laid before Parliament, at which time the Home Secretary can be questioned about its content and about the general administration of the Act, and to provide reasons should he have decided not to heed the Committee's advice.

In defining the Committee's role the Act lays down a provision which enshrines the basic principles and limitations of the legislation (Section 19 [2]):

> In its consideration of any matter the Committee shall have regard both to the legitimate requirements of science and industry and to the protection of animals against avoidable suffering and unnecessary use in scientific procedures.

In summary, the UK system provides all the elements identified in the introduction to this essay as essential to an effective system of control: a legislative framework, specified rules for the care and use of animals, and provision for ethical and scientific review. However, it will be argued in the Discussion section that fulfilment of the last criterion could be much improved within the UK control system.

EUROPEAN INITIATIVES

In 1986, a European Convention for the Protection of Vertebrate Animals Used for Experimental and Other Scientific Purposes was finalised (Council of Europe 1985). This document requires each of the 21 member countries of the Council of Europe, should they choose to ratify it, to pass national laws enforcing its rather weak provisions, though, unlike the EEC, the Council of Europe has no machinery whereby it can ensure this.

The Convention had a protracted and somewhat stormy history. Originally emanating from a Report adopted by the Councils Committee on Agriculture in 1970, it took a further 16 years for the final document, drawn up by an *ad hoc* Committee of Experts at Strasbourg, to emerge (Council of Europe 1985).

Weak though the provisions of the Convention are, it does provide for basic standards of laboratory animal protection in those countries where there is little or no national legislation. It contains minimum requirements for animal husbandry and care, for registration of establishments, and for proper marking and record keeping of animals (there is considerable concern about the possible use of stolen cats and dogs in some European countries), bans the use of stray cats and dogs, and requires the collection of statistical data.

Provisions for control over pain are extremely weak. The use of anaesthetics is encouraged but can be dispensed with if incompatible with the objects of the procedure, and even pain which is both severe and enduring (forbidden under the UK 1876 Act) is allowed under special authorisation. Perhaps the most important effects of the

Convention, with its exceptionally long history, lie not so much in the provisions of the final document but in the debate it stimulated in member countries to draw up national laws which often go beyond it, and in the stimulation it gave the Commission of the European Communities to draw up a Directive which is substantially stronger and which is enforceable in the 12 EEC member states.

The EEC Directive

In December 1985 the Commission put forward a proposal that the Community, in addition to the individual member states, should sign the European Convention. At the same time, in response to previous requests from the European Parliament it proposed that a Directive be drawn up. The Community became a signatory to the European Convention in February 1987.

The EEC Directive, On the Approximation of Laws, Regulations and Administrative Provisions of the Member States Regarding the Protection of Animals Used for Experimental and Other Scientific Purposes, was adopted by the Council of Ministers in November 1986, and in contrast to the Council of Europe Convention, its history was unusually short (Council of the European Communities 1986).

The main strengths of the Directive are that it requires adequate standards of husbandry and care; makes provision for the limitation of severe pain and distress; requires especial control by the Competent Authority in each member state of experiments liable to cause severe pain, including a weighing of the likely benefit against the severity; requires appropriate education and training of persons conducting or supervising experiments and those responsible for animal care; and makes provision for veterinary care and treatment, although it is optional to take veterinary advice on the treatment of an animal during an experiment.

The main weaknesses are that it places no obligation on member states to reduce animal experimentation, or to enforce the use of alternatives where they are available; it does not require the setting up of a Standing Committee in Brussels to monitor the working of the Directive; it makes no provision for the establishment of an Inspectorate in member states; the provision made for anaesthesia is inadequate in that it can be dispensed with even in the case of surgery, or when neuromuscular blocking agents are being used; there is no requirement for Competent Authorities to seek advice from animal-welfare bodies; and finally, the use of random-source animals is not prohibited, though the supply of cats and dogs is subjected to considerable control. Provisions for the control of pain are both weak and inconsistent. For example, Article 8.3 disallows the infliction of *severe* pain, while Article 10 forbids the re-use of an animal in a second experiment which entails severe pain or equivalent suffering.

The competence of the European Community extends only to matters covered by Article 100 of the Treaty of Rome which relates to trade between member states. For this reason, the Directive is restricted in its application to matters relating to the development and safety testing of products. The Commission has no authority to lay down rules concerning the conduct of research carried out for reasons such as the acquisition of scientific knowledge, education, or veterinary and medical application. However, on 14 November 1987 the Council of Ministers passed a Resolution agreeing that, in relation to experiments and procedures on animals falling outside the scope of the Directive, measures of control should be applied in member states which are at least as strict as those applied by it. In other words, the Directive will be used as a set of guiding principles for national laws covering all animal experimentation.

Against this background, many member states are currently drafting new national legislation or updating their current laws. It is clear that conditions between the different countries are varied. Some, such as the United Kingdom, have very elaborate control systems while others, such as Spain, currently have none at all. Table 10.1 provides a summary of the legislation in force or proposed at the time of writing.

CONTROLS FOCUSED AT THE INSTITUTIONAL LEVEL: THE USA

By and large, American researchers have maintained that animal-welfare considerations should be left to the conscience of individual experimenters, and they have regarded the UK control system (originally pressed for by British scientists and long accepted and even welcomed by them) as draconian. This thinking is reflected in the provisions of the US federal legislation, which was passed in response to public concern about the alleged use of stolen pets by laboratories after a family dog was traced to a laboratory and found dead there (Stevens 1978). There was also a perceived need to impose standards on animal husbandry and care, highlighted by a photographic exposé in *Life* magazine, in February 1966, of the squalid conditions in the facilities of one Maryland dealer.

Scope and enforcement of the US Animal Welfare Act

The US Animal Welfare Act of 1966 (amended in 1970, 1976 and 1985) applies to the transportation, sale, handling, care and treatment of

Table 10.1 Legislation in EEC Member States

Country	Controlling legislation	Administrative authority	Main controls
Belgium	Law on Animal Protection and Welfare (1986), chapter VIII	Ministry of Agriculture	Licences required for painful experiments. Breeders and suppliers must be authorised by Ministry. Records to be kept of source of dogs and cats. Laboratories using horses, dogs, cats, pigs, ruminants and primates must designate a veterinarian in charge of welfare. Users of other vertebrates must report it to Ministry. Leaders of projects to hold a degree in veterinary, medical, agricultural or biological sciences. Restriction of use of animals in teaching. Criteria laid down for qualification of technical staff. Central committee studying ethical issues. Inspection by State veterinary inspectors
Denmark	Law on Animal Experimentation (no. 382, 1987)	Ministry of Justice	Individual licensing of qualified persons for all experiments likely to cause suffering. Anaesthesia required for painful experiments but can be dispensed with. Procedures causing only minor pain not licensed but will be for cats, dogs and primates by 1989 and for all vertebrates by 1990. Restriction of use for teaching purposes. Animals must not be used where alternatives are available. Inspection by a Board (representing all interest groups) appointed by Ministry. The Board can withdraw licences. Detailed reports required by the Board

Country	Legislation	Authority	Description
Eire	Cruelty to Animals Act (1876)	Ministry of Health	Individual licences for experiments at registered places. Conditions attached relating to anaesthesia, limitation of pain, etc. Obligation to use anaesthesia waived if it frustrates the object of the experiment. Inspection by medical officers of Department of Health and veterinary officers of Department of Agriculture. Written records submitted annually to Minister of Health. Applicants for licences must testify that appropriate alternatives to animals are not available. Supplementary regulations will be required to implement the Directive, but not major changes
France	Decree no. 87–848 on animal experiments, Ministry of Agriculture (1987)	Interministerial Commission of the Ministries of Agriculture, Research and Higher Education	Individual authorisation from Department of Agriculture. Anaesthesia required unless it frustrates the object of experiment. No re-use of animals without anaesthetic. Inspection by Ministry of Agriculture veterinarians or Ministry of Public Health pharmacists. Decree was issued to fulfil requirements of EC Directive and will lay down standards for facilities and qualifications for persons carrying out research. It sets up a national advisory body on animal experimentation
Greece	Law 1197 concerning general protection of animals, art. 4 contains provisions on animal experiments (1981)	Ministry of Agriculture (aided by Consultative Committee with Veterinary Service of the Ministry)	Licences required for painful experiments. For surgical experiments anaesthesia must be administered by a veterinary surgeon. Surgical experiments restricted to graduates in medical, veterinary or biological sciences. No inspection. Much work needed to adopt controls which will fulfil new EEC requirements

Country	Controlling legislation	Administrative authority	Main controls
Federal Republic of Germany	Animal Protection Law (1972), parts 5 & 6, revised 1987	Ministry of Food, Agriculture and Health	Licences for qualified persons for each project issued by local authorities in the federal states, advised by local ethical committee with animal welfare, veterinary and scientific representatives. All possibly painful or injurious procedures on vertebrates are subject to licensing. Anaesthesia required for surgery but can be dispensed with if it frustrates the object. Experiments forbidden for weapons testing and, in principle, for testing tobacco products, washing agents and decorative cosmetics (exceptions may be granted). It must be convincingly shown that alternatives to use of animals are not available. Researchers required to weigh proposed benefits against ethical costs, though severe and enduring pain permitted if justified. Anaesthetics and analgesics required unless exceptional justification. Experiments authorised in advance except in fulfilment of legal or public health requirements. Each institution is required to appoint a scientifically qualified 'commissioner of animal welfare' to advise on animal use. Inspection by local authority veterinarians (not specialists in animal experimentation). Cats and dogs to be purpose-bred, though exemptions allowed
Italy	Animal Protection Law no. 615 (1941)	Ministries for Health and Culture	Experiments performed by named, qualified individuals in authorised institutes. Responsibility rests with

		director. Anaesthesia required unless it frustrates the object of the experiment. Annual report required. Inspection by medical and veterinary officers of provincial health authorities. Services understaffed, changes required to meet EEC requirements
The Netherlands	Experiments on Animals Act (1977). Order in Council (1980) being implemented	Ministry of Public Health; Veterinary Public Health Inspectorate— animal experimentation department	Retrospective licences to institutions (no project assessment) are compulsory for experiments on vertebrates causing pain or suffering. Project leaders must be trained. Named person responsible for animal welfare in each institute: these are veterinarians, doctors or biologists who have had specific training. Neither they nor inspectors can stop experiments, but can impose restrictions. Animals in pain must be euthanased once experiment is completed, but anaesthesia can be dispensed with if it frustrates the object of the experiment. Alternatives to animals must be used where available. No cats, dogs, horses or primates used if other species will suffice. Source of dogs and cats recorded. Strict rules on supply but exemptions allowed. Inspection by state and regional veterinary inspectors. Annual return with detailed statistics required. Central advisory committee includes laboratory animal scientists and experts on animal welfare. Annual report produced. New controls being phased in, and detailed provisions on husbandry and care are under consideration. Ethical committees will be compulsory after 1989

Country	Controlling legislation	Administrative authority	Main controls
Grand Duchy of Luxembourg	Law for the Protection and Welfare of Animals (1983)	Ministry of Agriculture	Licences issued by the Ministry of Health. Inspection by Ministry of Agriculture veterinarians. Provisions similar to the Animal Protection Law of the Federal Republic of Germany
Portugal	No current legislation. Legislation required to comply with EEC Directive		—
Spain	No current legislation. Royal Decree 3181 (1980) requires authorisation		Royal Decree requires authorisation for capture of protected wild species. Regional community governments are autonomous. In 1986 an interministerial committee was set up giving the Ministry of Agriculture powers to lay down control mechanisms which they might follow. A Royal Decree is in preparation.
The United Kingdom	see text		

certain species of research animals by directing the Secretary of Agriculture (hereafter Secretary) to promulgate Regulations setting down humane standards of care. Research facilities are required to be registered by the US Department of Agriculture (USDA) and are inspected by it.

The Act was never intended to control actual research practices and has failed even to implement the original intention of Congress to provide for adequate animal care (Hampson 1987 a, b). This has been forcibly demonstrated not only by the dramatic cases brought to light by animal liberationist raids on facilities over the last few years but also by evidence in USDA inspectors' own reports which have documented dramatic non-compliance even with the minimal provisions of the Act.

The Animal Welfare Institute (AWI), Washington, DC, drawing on these reports, which are available through the US Freedom of Information Act, has documented abuse or neglect in 82.7% of the facilities investigated in a four-year period and the fact that less than one-fifth were in compliance even with the basic provisions of the Act (AWI 1985). Their documentation describes not only neglected animals in badly maintained facilities but also poor scientific procedure such as a failure to wash hoods between carcinogenicity tests, resulting in animals being unintentionally submitted to carcinogen cocktails. In recent years a number of cases investigated by USDA, and by the National Institutes of Health, of federally funded facilities have revealed, among other things, grossly filthy conditions and inadequate veterinary care in some US facilities and have led to withdrawal of federal research funds and fines by USDA under the Act.

Enforcement of the Animal Welfare Act is ostensibly by the Animal and Plant Health Inspection Service (APHIS) of the USDA. The role of this body has primarily been livestock disease control and historically the Service has had neither the motivation, the budget nor the specialist training needed to carry out the task of monitoring the USA's massive research effort.

Though no accurate statistics are available the number of animals used per year in the USA, in over 1200 facilities scattered over 3.6 million square miles, had been variously estimated at between 25 and 70 million (Office of Technology Assessment 1986). Official sources now indicate that the figure is around 25–30 million, more than 80% of which are rats, mice and birds, which are not covered by the Act. Horses and farm animals are likewise excluded.

This major omission was partly due to the historical reasons for the Act's passage, partly due to pressure from the scientific community, and partly due to lack of resources. Facilities using only these species are outwith the Act and are not open to federal inspection by USDA.

Though the 1970 amendments broadened the definition of 'animal' under the Act to include any warm-blooded vertebrate determined by the Secretary as being used in research and testing, Regulations have continued to exclude rats, mice, birds, horses and farm animals. Moreover, facilities not using 'substantial numbers' of animals may also be exempted by the Secretary.

A recent report by the Office of Technology Assessment (1986) takes the view that exclusion of these species appears to frustrate the policy sought by Congress in the 1970 amendment and was beyond the statutory authority vested in the Secretary of Agriculture.

New controls over animal care and use

Various amendments to the Act have defined more clearly the standards to be promulgated by the Secretary of Agriculture in Regulations which implement it. The most important of these was that there should be adequate veterinary care 'including the appropriate use of anaesthetic, analgesic or tranquillising drugs when such use would be proper in the opinion of the attending veterinarian' of the research facilities. These provisions were the first hint of interference with research practices, though at the time they were never properly defined nor enforced, and attending veterinarians would frequently sign the required annual report to USDA, stating that use of such drugs had not been necessary. Thus the reporting system failed to provide Congress with the measure of public accountability it had sought concerning control over pain and distress (Solomon and Lovenheim 1982).

Much more significant were the 1985 Amendments which were a response to increasing public concern and sought to rectify some of the Act's major inadequacies and omissions. These were implemented through the Improved Standards for the Laboratory Animals Act, public law 99–158, which enacted a number of proposals put forward in the Dole and Brown private Bills which had undergone much discussion in the Senate and Congress, respectively. These proposed a mechanism whereby control could be exerted at an institutional level and overseen by the federal inspection machinery.

The amendments required the establishment, at each institution, of an Animal Care and Use Committee (ACUC), its membership to include the laboratory veterinarian and at least one non-institutional member. Its duties are regularly to review and inspect animals and procedures and to file an inspection report with the USDA. Laboratory primates are to have an environment commensurate with their psychological well-being, and dogs must be given exercise. Adequate standards of veterinary care and the use of anaesthetics, analgesics and tranquillisers are promulgated. Paralytic drugs are not to be used without adequate anaesthesia and multiple survival surgery is to be strictly limited.

Alternatives to the use of animals must be given consideration and an information centre for data exchange on alternatives is to be established at the National Agricultural Library. Each facility should provide training in humane experimental technique. New civil penalties are provided for violation of the Act, and the role of USDA Inspectors expanded to include at least an annual inspection of each facility. The Secretary of Agriculture is to promulgate rules to implement these provisions, after due consultation.

These proposed improvements to the control system will relate at least as much to the husbandry and care of animals as to the actual conduct of research and inspectors are expressly forbidden to interfere with the progress of research during their visits.

Controls exerted by the funding agencies

The system of control in the USA is actually more complex than is laid down under the Animal Welfare Act. The Act, which applies to all registered facilities, puts into practice provisions similar to those which were already applied contractually to institutions receiving federal research funds, which accounts for a large percentage of research conducted in the United States. Such institutions are required to comply not only with the provisions of the Act but also with guidelines issued under the National Institutes of Health (NIH) to all facilities receiving NIH grants, as required by Public Health Service (PHS) policies.

The status of the NIH Guide (NIH 1985a) was enhanced, and its provisions strengthened, in 1985 by the passage of the Research Extension Act (NIH Reauthorisation Bill) which required detailed Guidelines to be established for care and treatment of animals in *all* research facilities, not just those receiving federal funds (NIH 1985b).

These changes required the setting up of Animal Care and Use Committees (ACUCs) with involvement in all aspects of the research programme; designation of clear lines of authority in the institution with two named officials, one responsible for the institution's animal programme and the other a veterinarian; detailed written information of the animal care and use programme to be filed with the NIH Office for Protection from Research Risks (OPRR) which administers these controls; and a review by the ACUCs of all aspects of applications for research funding which relate to the care and use of all animals (including rats and mice).

Allocation of research grants is dependent upon this documentation, and the 1985 amendments mandated the seizure of federal funds from institutions which violated the Guidelines. NIH carries out its own inspections of institutions receiving its grants. In recent years, since an increasing number of animal-welfare problems have come to light, the frequency of unannounced visits and their thoroughness have increased. Before major violations were exposed, NIH had shown little interest in enforcing its own rules and had never seized funds from a facility. It now monitors not only compliance with the Guide but also functioning of the ACUCs which are now in place in all institutions.

PHS policy was further strengthened in 1987, requiring the ACUCs to inspect facilities, including satellites, twice annually, the filing of minority as well as full committee reports with OPRR and an explanation to be given by the facility of the training programme for researchers and animal care personnel on humane practices. Both NIH and OPRR have been running a series of workshops on implementation of the new PHS policies.

Like the Animal Welfare Act, NIH Guidelines do not specifically interfere with the conduct of research. Use of anaesthesia and pain-relieving techniques is at the discretion of the experimenter. There is no restriction on the types of research techniques which may generally be used (with some exceptions, such as the prohibition of the use of neuromuscular blockers without anaesthesia) and no species-specific detail about behavioural and environmental needs is given in the Guide.

In addition to these controls a proportion of institutions subscribe to the voluntary accreditation scheme offered by the American Association for Accreditation of Laboratory Animal Care (AAALAC) which, it has been claimed, offers the guarantee that accredited facilities meet the highest standards of husbandry and care, though a number of institutions cited for non-compliance with the Animal Welfare Act have had AAALAC accreditation.

Mechanism of the controls; coordination and controversy

Institutions carrying out toxicity tests on behalf of government agencies, such as the Food and Drugs Administration (FDA) and the Environmental Protection Agency (EPA), are further required to comply with various standards of animal care and treatment set down in the Good Laboratory Practices Act of 1978. These were a response to FDA concerns about shoddy experimental methodology and inadequately maintained facilities.

In order to eliminate confusion and inconsistencies which may arise when institutions are conforming to differing standards laid down either by the USDA under the Animal Welfare Act or by PHS policy, there is a Memorandum of Understanding between the FDA, APHIS and NIH which exists to ensure that, when problems arise, only one agency will arrive to inspect the facility. This empowers USDA inspectors to report violations of PHS policy to NIH even where such violation relates to areas or animals not covered by the Animal Welfare Act. Procedures for consultation between the Secretary, OPRR and APHIS are intended to iron out policy inconsistencies, though currently minor problems still exist.

Presently, a heated debate is raging in the USA concerning the proposed Regulations for implementing the 1985 amendments to the Act, which were issued by the USDA in March 1987 (USDA 1987). The Department proposes that the Annual Report of a research facility, which is required to be certified by key members of the facility and filed with APHIS, should demonstrate evidence of acceptable standards of animal care and treatment. The ACUC is required to approve protocols for all research involving animals in significant but unavoidable pain or distress. Procedures involving severe pain or distress or death may be approved provided that the experimenter asserts that there is no alternative method of gaining the scientific information required by the experiment.

The most controversial proposal is that the ACUC can approve exceptions to the standards laid down. In addition to inspectorial and protocol review duties the ACUCs are to be charged with the task of writing and implementing (in consultation with the veterinarian) the new provisions for exercising dogs and for meeting the behavioural and psychological needs of primates. These Regulations are expected to be published in 1989.

The basic challenge from the research community revolves around the proposal to regulate conduct of research by requiring ACUCs to review protocols. The research lobby claims that this measure exceeds authority given to USDA under the law. Though the law does not explicitly require protocol review, it is clear that it was the intention of Congress that this should be one of the Committee's primary duties. They also argue the bureaucratic unwieldiness of having to work under several sets of controls at once, and claim that the workloads of the committees are so great as to be impracticable.

The more extreme faction is opposing the Regulations almost in their entirety, and sees this attempt by Congress to ensure that minimisation of animal suffering goes hand in hand with good science as an unacceptable intrusion by government into free scientific enquiry. Meanwhile, the animal protection lobby has made out a strong case showing that the proposed Rules are in compliance with the Act and with the intention of Congress behind it. It is also campaigning vigorously to have the provisions of the Act extended to cover mice, rats and farm animals, arguing that their exclusion is an oversight which was never intended by Congress.

DISCUSSION: CAPABILITIES AND LIMITATIONS OF LEGISLATIVE SYSTEMS

To what extent can legislative systems hope to fulfil the requirements set out in the introduction to this essay?

- To define legitimate purposes for which animals may be used.
- To exert control over allowable levels of pain and other distress.
- To provide for inspection of facilities and procedures.
- To ensure humane standards of animal husbandry and care.
- To ensure public accountability.

To answer this question adequately would require a major review far beyond the scope of this essay. However, some comparisons can be drawn between the two main types of system outlined here—control by central licensing and control by institutional committees. The UK and USA have been chosen for detailed consideration because they have perhaps the best-known and most widely debated systems at the current time.

Control over purpose

As already indicated, most legislative systems currently in force do not address the question of purpose other than in a very general way. Most which have any form of prior authorisation do require the applicant to show that the work is important and necessary and that there is no viable non-sentient alternative available which could provide the same information. These very general requirements may serve to raise consciousness in individual researchers, though the possibility remains that they may be quickly dismissed by a few sentences on an application form.

The clear responsibility placed on the Home Secretary under the new UK Animals (Scientific Procedures) Act 1986 to weigh the cost to the animal against the proposed benefits of the research (Section 5) (see also the earlier section in this essay on the UK 1986 Act) is a unique provision in animal experimentation law and one which was hard won by the reforming animal-welfare movement. It does at least provide the possibility that the purpose of a project will be scrutinised and weighed against the degree of suffering to which the animals are likely to be subjected. Yet this scrutiny will not meet the requirements envisaged by reformers unless it addresses the fundamental ethical question of whether a research project should be conducted at all. Projects are assessed primarily by Home Office Inspectors who are not ethicists and they would not usually see it as their role to advise the Secretary of State against allowing a piece of research.

In performing its cost–benefit assessment the Animal Procedures Committee (APC) *could* advise the Secretary of State to proscribe certain kinds of research on the grounds that the general projected benefit cannot justify the animal suffering involved. Whether or not the Committee will adopt this line of argument, indeed whether it will consider it to be within its remit, has yet to be seen.

Could an institutional review committee do a better job at cost–benefit assessment? The answer is probably that it ought to, since such a committee is meant to have represented on it a range of interests and expertise, and such a range is necessary for the proper conduct of this kind of analysis. However, US Animal Care and Use Committees (ACUCs) do not use the name of 'ethical committees' and do not see themselves as such. As we have seen, they are operating within a framework which is hardly designed to promote ethical review. It is too early to tell how well the new system might work in the USA. At least one legal review has suggested that federal regulations should require that protocols be approved only when it can clearly be shown that the benefits of the research 'clearly outweigh' the costs to the subjects (Dresser 1986). ACUCs are a long way from attaining, or even perceiving, such a goal.

Control over pain and suffering

The UK system imposes controls over permissible levels of pain and suffering which are arguably the strictest found anywhere in the world. Unlike the US system, whose objective was never to interfere significantly with the conduct of research, the avowed intention of the new UK legislation, as explained by the Minister during its passage, is a reduction in the number of animals used and in their suffering (see essay by Hollands in this volume).

Project licensing with its severity banding should ensure a deeper scrutiny of protocols at the licensing stage, enabling the Inspectorate to impose conditions upon those projects most likely to result in considerable pain and suffering. Such conditions will include specification of the type of anaesthesia to be used, the method of euthanasia and the use of analgesics. The whole debate on these topics has opened up, largely as a result of discussion between veterinarians, researchers and the Inspectorate during the implementation of the new controls. In this sense, new legislation brings with it improvements to laboratory animal welfare which are not immediately apparent in the content of the controls themselves. It is now acknowledged, for example, that some methods of anaesthesia, such as ether for small rodents, are not desirable from either a humane or a safety perspective. Such initiatives should lead, hopefully, to more research into refinements and better training of laboratory personnel in the application of humane techniques.

They also are leading to research into definition and recognition of pain and suffering. While definition will always be subjective, veterinarians nonetheless are involved daily with recognising and treating suffering. Such a rationale underlies the development of scoring schemes for levels of suffering according to a number of behavioural and other criteria (Association of Veterinary Teachers and Research Workers 1986, Morton and Griffiths 1985). These schemes can form a basis for pain alleviation, for example, with analgesia, and they are being tested by a number of research workers both in the UK and elsewhere (Beynen *et al.* 1987a, b).

The UK termination condition specifies that no animal must be allowed to suffer pain or other distress which can be described as severe and this requirement, together with the new authority now vested in the named veterinarian and person responsible for day-to-day care, should lead to more serious attempts to define what is meant by such terms than were ever made under the 1876 Act. On-site availability of caring personnel with statutory authority to act in the interests of animals is essential to the effective working of *any* control system.

In some countries schemes have been drawn up to categorise procedures according to level of severity. The Scientists' Center for Animal Welfare (SCAW) in Washington has drawn up such a categorisation with the aim of helping ACUCs carry out protocol review (SCAW 1986). While it is certainly useful to flag projects which may require more detailed scrutiny by ranking types of procedure according to degree of severity, the UK Home Office has remained opposed to incorporating any such categorisation into its project licensing system. It takes the view that it is better not to cast certain kinds of procedure in stone by giving them a category, but rather to let a consensus emerge naturally within the Inspectorate through the scrutiny of individual project licences.

The USDA Regulations describe three levels of pain or discomfort: routine procedures requiring no anaesthetic or analgesic; levels of pain and distress which require amelioration; and a third category, the most severe, which require approval of the ACUCs. They do not spell out clear standards for post-procedural pain relief, though these do appear in the NIH Guide. Most importantly, though specific procedures, such as the infliction of severe trauma on anaesthetised animals, use of neuromuscular blockers without anaesthesia and methods of euthanasia proscribed in the guidelines of the American Veterinary Association, are banned by national policies, there is no general principle in the legislation placing a limit on the degree of pain or distress which can be caused where the use of pain relief would frustrate the object of the experiment.

In the USA, the committees ought to be given a clear brief from the national authorities to review protocols from the perspective of minimising pain and distress by use of appropriate methods of analgesia, anaesthesia and euthanasia, and there should be a pre-scribed limit at which it is required that animals be humanely killed. They should consistently adopt specific policies on refinement of specific procedures such as the use of Freud's complete adjuvant, tests with death endpoints, prolonged physical restraint and the infliction of trauma. Through the recommendation of alternative techniques or refinements they should build up a set of guidelines on appropriate means of carrying out particular kinds of technique.

Monitoring and inspection

It is arguable that the UK Inspectorate, which has been in force for over a century, is the most effective in the world. It is made up not only of veterinarians but also of medics, since the Home Office has argued that this facilitates scrutiny of licences and assessment of the impor-tance of some research work. For a number of years the practice of medics and vets working together in their local areas has enabled a sharing of this range of expertise at the local level. Under the new legislation the strength of the Inspectorate has been increased to 25 and each Inspector has begun to specialise in a particular research area such as cosmetics testing or behavioural research. This expertise is shared at the central level.

In the USA, as already outlined, there has been a striking failure of the system to ensure inspection and enforcement of even the most basic animal protection provisions contained in the Animal Welfare Act, prior to its recent amendment. The primary reasons for this deficiency have been lack of specialist training and motivation on the part of USDA Inspectors and a totally inadequate budget from Congress to do the job. This is now changing and APHIS Inspectors are taking a much greater interest in their role under this legislation.

Under the proposed regulations a new system will be introduced requiring correction of deficiencies within 30 days of their having been reported to the APHIS. This is opposed by many researchers who complain that many deficiencies will take longer than this to correct, though there is room for individual negotiations. In addition, NIH has recently been making unannounced visits in response to complaints or perceived problems, though routine site inspections are still a rare occurrence. On such visits the aim is to ensure that the machinery exists at the institutional level for identifying and correcting problems.

To be effective at the institutional level, the ACUCs must have the machinery to monitor what takes place in the laboratory itself. If, for example, the committee makes recommendations for refinement of a protocol, it has the means to ensure that its recommendations are carried out. It also familiarises the committee with the actual situation in the facility and enables dialogue between it and the laboratory personnel. Such dialogue, as the UK Inspectorate has shown, is essential to the effective implementation of a control system which can come to be seen as helpful rather than adversarial.

For this to work well the institution must ensure that the committee operates through accepted lines of authority. The veterinarian should have an overview brief and should form a functional link between the committee and animal care personnel, ensuring that skilled staff with innovative ideas have an official means of expressing their views. If institutions can manage to operate this kind of machinery effectively there will be more openness in the whole administrative system and a greater awareness of issues amongst all those who have to work within it.

Laboratory animal husbandry and care

No legislative control system, however good on paper, will work unless it can be effectively translated into operation in the laboratory on a daily basis. One of the great strengths of the new UK system is the responsibility vested in the named veterinarian and person responsible for day-to-day care. Because of the structure and long history of control in the UK, these duties are carried out in close co-operation with researchers and the Home Office Inspectorate. This means that the named persons can do far more than perform their basic functions as outlined under the Act; they are in a position to promote innovations in the care of animals and to assist with the development of refinements to husbandry and experimental techniques.

One indication that they are already beginning to perceive such a role for themselves is the fact that symposia and workshops on animal care and alleviation of pain in the UK are increasingly heavily attended by veterinarians and senior technicians, and often many of the papers are given by them. Such people were not nearly so much in evidence a few years ago. This contrasts to some considerable degree with the current situation in the USA. In many institutions veterinarians have not yet settled into a comfortable role within the system and have often been in conflict with members of the institutional Animal Care and Use Committees on such matters as exercise for dogs.

Professional veterinary bodies have complained, for example, that the cost of implementing the proposed regulations in respect of dogs and primates will be in excess of two million dollars. On the whole, in the USA, veterinarians do not want the direct responsibilities which the new system would impose upon them, and yet they are an essential link between the laboratory and the ACUC if the system is to work at all.

In any system, guidelines issued under the legislation play a major role in promoting improved animal welfare. Indeed it is usually the guidelines, rather than the law itself, which translate intention into practice and elaborate the specific responsibilities of laboratory personnel. For example, under the UK Act the Home Secretary has now issued a Code of Practice which is based broadly on the detailed Guidelines on Husbandry and Care recently drawn up by an expert working group of the Royal Society (RS) and the Universities Federation for Animal Welfare (UFAW). These were published in July 1987 and further sections, for example on recognition and alleviation of pain, are currently in production.

There is a danger that rigid guidelines, which require institutions to spend large amounts of money on increasing the size of their cages, could be counter-productive, because they stifle real innovations which might better address the needs of the animal. For example, a number of UK institutions are now housing ferrets in groups in rooms, as cats have usually been housed for many years, and some institutions are beginning to experiment with a similar system for rabbits. As the RS/UFAW Guidelines stress, such changes are perhaps even more pressing for non-human primates, whose ethological needs dictate that they be housed in social groups.

Such changes may require an increase in laboratory personnel, and they will certainly require a high standard of training in handling and care and a high level of commitment to improving the well-being of the animals as a priority. The UK government has now recognised the need for training animal carers, at least in the case of non-human primates, and in response to a recent report submitted to the Home Secretary (FRAME/CRAE 1987) has undertaken to look into the question of providing funds for training in the care of these animals.

Training is perhaps the single most important factor in any system of control, but insufficient attention has been paid to it. The UK government has, in the past, done very little to promote training and there are no official requirements for standards of training, though the granting of a personal licence depends upon the Inspector's assessment of the researcher's level of competence, and restrictions and conditions may be applied to the licence on the basis of this assessment. In this sense there is perhaps more control over competence than in any other system, but the only country which actually requires formal training by a prescribed course, both for researchers and for laboratory veterinarians, is the Netherlands (see Table 10.1).

In the UK, animal technicians are trained to a high level through courses offered by the Institute of Animal Technology, enabling them to attain recognised standards. Researchers are often trained in-house and standards are variable. A number of institutions are now offering short courses for researchers and technicians, some of which are general and contain basic practical instruction in handling, simple techniques, housing and sometimes ethical concepts and alternatives. Some more specialised courses are also offered relating to care and pain relief in particular species. Training for laboratory animal veterinarians is also often in-house. A two-year MSc course is offered by the Royal Veterinary College and the British Laboratory Animal Veterinary Association runs several courses. The Royal College of Veterinary Surgeons offers examination for a diploma in laboratory animal science for veterinarians with adequate specialist training.

The new provisions in the US Animal Welfare Act require that each institution provide adequate training for all its personnel. The training programme is to be reviewed by the ACUCs. Research organisations have argued that this is not a suitable task for the committees to undertake and that this responsibility should rest with the institution itself. However, in many institutions the committee is well qualified to undertake the task. Its members include a range of researchers and animal care specialists and sometimes ethicists. The committee is also involved in institutional policy and therefore is familiar with the institution's individual needs. If information can be exchanged between committees through practical workshops, this would seem to be an excellent way to promote training and to achieve some degree of consistency.

As in the UK in the past, though the situation is now changing, some members of the US research community are averse to training schemes, arguing that training in ethics is irrelevant and practical training unnecessary. They expect to 'pick up' the necessary skills to handle animals in much the same way they would learn to use any other technical equipment. Many argue that they do not have the time to undergo training courses.

Clearly such arguments hold very little water in the face of evidence often given at meetings by highly trained technicians, regarding the improper handling and sloppy procedures which they have to put up with. Well-trained technicians have a wealth of information to offer researchers. Their membership on the ACUCs ought to be compulsory, and they ought to be listened to.

NIH has recently instituted a four-hour course for investigators to introduce them to the topics of humane treatment, public interest in the area, investigator responsibilities and opportunities which exist for training. As researchers come to realise more and more their responsibilities on the ACUCs, resistance to training can be expected to fade.

The real needs are for researchers to obtain practical hands-on training, which is both costly and time-consuming. Yet without it there is little point in elaborating control systems that require them to reduce animal suffering. Just as essential is some basic introduction to ethical evaluation. Rarely are researchers introduced to such concepts. Many have never asked themselves 'Is the proposed benefit of this work worth the cost to the animals?' and they have learnt no specialised skills which would enable them to make such an evaluation. However, with intensifying debate, many researchers have begun to question their work more deeply. What they need now is guidance.

Ensuring public accountability

As we have seen, the new UK legislation seeks some measure of public accountability, through the publication of annual statistics and an annual report of the Animal Procedures Committee, which can be regarded as a public watchdog. The public can raise questions on these documents and other aspects of the control system through the Parliamentary process, though it has no direct access to details of the workings of the control system since the UK has no Freedom of Information Act, and individuals have no right of entry to laboratories. Moreover, the public has no direct input into the control system, in the way that the appointment of lay-members to institutional committees allows; they must rely on the lay and professional animal-welfare members of the Animal Procedures Committee to represent their interests. These members do, in fact, sit on the Committee as individuals, and the more radical animal protection groups who wish to see faster and more swingeing reforms than those effected in the legislation can rightly claim that there is no room in the system for the representation of their interests.

The USA relies on an institutional control system to provide public accountability since central administration, involving several agencies, is complex, and it would be difficult to set up a national advisory body. However, the setting up of some sort of national panel as a forum for discussion would be extremely useful at the present time. Its members could be actively involved in institutional committees and there would be opportunity for researchers and animal-care personnel to discuss the functioning of these committees and to produce guidelines. This would also facilitate co-ordination between the committees and would help to achieve some degree of consistency. The committees could also discuss ethical issues and matters of general policy such as the sorts of procedures that should be proscribed.

One requirement for the assurance of public accountability through such a system is that the committees be satisfactorily composed, and it is clear that they should have at least one member whose primary position is to identify with the research animal. This is not anthropomorphic; it is a straight analogy with the lay member of an Institutional Review Board (IRB) (the US equivalent of UK clinical experimentation ethics committees), who is a person who might him/herself be subjected to the experiments under review. In this sense the analogy with IRBs, on which the American ACUCs are supposed to be modelled, has been somewhat over-stretched. In the case of human experimentation the research subject is capable of consent. Animals have no real moral status in our society and they cannot consent to procedures which can be highly invasive and which may result in suffering and death.

If ACUCs are in any way to satisfy the requirements of public accountability their composition should reflect the concerns of society as a whole and not just research interests. The Scientists' Center for Animal Welfare has drawn up consensus recommendations which state that the chairperson should be someone other than the animal faculty director or the attending veterinarian; a senior administrator or other faculty person might be more appropriate. It recommends that persons representing community concerns should be present, and in addition there could be representatives from other scholarly disciplines such as the humanities, ethics and the law (SCAW 1986).

There is no doubt that, when they work well, institutional committees can do a great deal to promote the effective working of a control system and to assure public accountability. It could well be argued that they would be a useful adjunct to the centrally operated control system in the UK. They need not, as is so often argued, simply add to the bureaucracy since they ought to cut through it by helping to design projects for scrutiny by the Inspectorate and saving time at an early stage by flagging areas of difficulty and potential for refinement. Such committees would also go some way towards meeting complaints that the public and animal-welfare groups have no input into the UK system of control.

CONCLUSION

Modern legislative initiatives are seeking to respond to the increasing demand that the benefits of research outweigh the costs to the animals involved. The balance can be effected either by central administration, as through the UK project licence, or at the institutional level, through Animal Care and Use Committees. Both systems are beset with problems, but either *could* work. It would require only goodwill on the part of all those administering legislation and those working under it to make changes which would really benefit animals. It would be necessary to put this goal as high on the list of priorities as reaping the benefits of research.

We are a long way from attaining that goal in any country, but the general debate and the initiatives in animal care and use which have come out of attempts to strengthen legislation have brought us a good deal nearer to it. Legislation which would prohibit animal experiments is hardly conceivable in our society; profound societal changes would have to precede this. It is unlikely that we can even achieve legislation which would ban trivial and questionable research, or remove all pain and suffering from experiments. It is for this reason that some animal-rights groups have chosen to dissociate themselves from the legislative debate. They believe that it is better to have no reform at all than marginal improvements which will only make it more difficult to achieve radical reform.

There is little historical support for these claims. The enfranchise-ment of women took place in two stages, first for women over 30 and much later for women aged 21–29. There is abundant evidence that moderate efforts do lead to meaningful reform, and from the animals' point of view it is surely better to be housed in a compound than a cage and to be given a pain-killer than not.

It is possible to see already that legislative reforms *are* making a difference. In the UK the severity banding and project licensing system is having a real effect in reducing severity, largely as a result of the effort being put into the exercise by inspectors, veterinarians and researchers. No-one knows, for example, how much pain and distress is caused by many routine procedures. Now there is a *need* to know in order to categorise licence applications and this is leading to studies to find objective measurements. In the USA, changes in the Animal Welfare Act and public policies are having a major impact on the way institutions make judgements about animal use. The last few years have seen an ever-increasing number of international symposia dealing with definition, recognition and relief of laboratory animal pain.

Over the last few years, the animal-rights debate has highlighted the serious ethical issues which underlie emotional empathy with animals. The price we pay for our material gains from animal research is the loss of a sense of kinship with the rest of sentient creation. Thus we have stifled the possibility of fostering a consistent moral and compassionate attitude towards the other creatures with which we share our planet.

It is conceivable that, in the long term, regulation of animal experimentation could incorporate protection of animal rights as well as utilitarian principles. Thus it might one day become politically defensible to protect animals in the same way that society protects vulnerable humans, such as the mentally defective, by appointing legal guardians and by limiting the degree of risk to which they can be subjected.

Such a possibility is far into the future, but the door is opened to it by the debate which legislative reform has now started. We see glimmers of it in initiatives such as the recent convening of a three-year study by the Institute of Medical Ethics in the UK to investigate the ethics of using animals in research. When we look at the wider context of the use of animals in research, clearly we can see legislation as a beginning and not as an end.

REFERENCES

Animals (Scientific Procedures) Act (1986), Eliz. II, c 14, HMSO, London

Association of Veterinary Teachers and Research Workers (1986). Guidelines for the recognition and assessment of pain in animals, *Vet. Rec.*, **118**, 334–8

AWI (Animal Welfare Institute) (1985). *Beyond the Laboratory Door*, Animal Welfare Institute, Washington, DC

Beynen, A. C., Baumans, V., Bertens, A. P. M., Havenaar, R., Hesp, A. P. M. and van Zutphen, L. F. M. (1987a). Assessment of discomfort in gallstone-bearing mice: a practical example of the problems encountered in an attempt to recognise discomfort in laboratory animals, *Lab. Anim.*, **21**, 35–42

Beynen, A. C., Baumans, V., Haas, J. W. M., van Hellemond, K. K., Stafleu, F. R. and van Tintelen, G. (1987b). Assessment of discomfort by orbital puncture in rats. In Beynen, A. C. and Solleveld, M. A. (eds), *New Developments in Biosciences: Their Implications for Laboratory Animal Science*, Martinus Nijhoff, The Hague, pp. 431–6

CIOMS (Council for International Organisations of Medical Sciences) (1985). *International Guiding Principles for Biomedical Research Involving Animals*, CIOMS, Geneva

Clark, S. (1977). *The Moral Status of Animals*, Clarendon, Oxford

Council of Europe Committee *ad hoc* Protection des Animaux (1985). *European Convention for the Protection of Vertebrate Animals Used for Experimental and Scientific Purposes*, Council of Europe, Strasbourg

Council of the European Communities (1986). *Council Directive on the Approximation of Laws, Regulations and Administrative Provisions of the Member States Regarding the Protection of Animals Used for Experimental and Other Scientific Purposes*, Official Journal of the European Economic Communities, Series L, no. 358, pp. 1–28

Dresser, R. (1986). Research on animals: values, politics and regulatory reform, *S. Carolina Law Rev.*, **58**, 1147–201

FRAME/CRAE (Fund for the Replacement of Animals in Medical Experiments and the Committee for the Reform of Animal Experimentation) (1987). *The Use of Non-Human Primates as Laboratory Animals in Great Britain*, FRAME/CRAE, Nottingham

Hampson, J. E. (1987a). In Tuffery, A. A. (ed.), *Laboratory Animals: An Introduction for New Experimenters*, John Wiley, Chichester, pp. 21–52

Hampson, J. E. (1987b). In Rupke, N. A. (ed.), *Vivisection in Historical Perspective*, Croom Helm, London, pp. 314–39

Hansard (1986). House of Commons Reports, Vol. 96, no. 97, col. 98, 12 April

Home Office (1986). *Home Office Guidance on the Operation of the Animals (Scientific Procedures) Act 1986*, London

Midgley, M. (1983). *Animals and Why They Matter*, Penguin, Harmondsworth

Morton, D. B. and Griffiths, P. H. M. (1985). Guidelines on the recognition of pain, distress and discomfort in experimental animals, *Vet. Rec.*, **116**, 431–6

NIH (National Institutes of Health) (1985a). *NIH Guide for Grants and Contracts*, US Department of Health and Human Services, US Government Printing Office, Washington, DC

NIH (1985b). *Guide for the Care and Use of Laboratory Animals*, US Department of Health and Human Services, NIH, Bethesda, MD

Office of Technology Assessment (1986). *Alternatives to Animal Use in Research Testing and Education*, US Congress Office of Technology Assessment, Washington, DC

Regan, T. (1983). *The Case for Animal Rights*, Routledge & Kegan Paul, London

Rollin, B. (1981). *Animal Rights and Human Morality*, Prometheus Books, New York

Rowsell, H. (1987). In Orlans, F. B., Simmonds, R. C. and Dodds, W. J. (eds), Effective Animal Care and Use Committees, *Lab. Anim. Sci.*, special issue, 24–7

SCAW (Scientists' Center for Animal Welfare) (1986). Consensus recommendations on effective institutional animal care and use committees. In Orlans, F. B., Simmonds, R. C. and Dodds, W. J. (eds), Effective Animal Care and Use Committees, *Lab. Anim. Sci.*, special issue, 11–13

Singer, P. (1975). *Animal Liberation: A New Ethics for Our Treatment of Animals*, Avon Books, New York

Singer, P. (1985). *In Defence of Animals*, Basil Blackwell, Oxford

Solomon, M. and Lovenheim, P. C. (1982). Reporting requirements under the Animal Welfare Act: their inadequacies and the public's right to know, *Int. J. Study Anim. Prob.*, **3**, 210–18

Stevens, C. (1978). In Leavitt, E. S. (ed.), *Animals and Their Legal Rights*, Animal Welfare Institute, Washington, DC, pp. 46–58

USDA (United States Department of Agriculture) Animal and Plant Health Inspection Service (1987). Animal welfare: proposed rules, *Fed. Register*, **52**, no. 61, 31 March

Appendix
Current Legislative
Initiatives in the USA

Holly Hazard*

INTRODUCTION

One fundamental difference exists between the US regulation of animal experiments and regulation in Western Europe. In most of Europe legislation of one sort or another has managed to slip into law that, in some small way, places restrictions on the very act of experimentation. In some countries, some uses have been found to be unjustified. In Denmark, for example, the use of animals in teaching is legally restricted. In others, restrictions may be imposed for military experiments or for the testing of cosmetics. Many countries have taken steps to regulate pain during experimentation. Still others have placed restrictions on the kinds of animals acceptable for experiment protocols. In the USA however, no animal is too sentient and no experiment is too silly, painful or duplicative for the government to restrict its implementation. Legislators have compartmentalized compassion into care before or after experimentation versus care during experimentation. Even the most thoughtful or caring of our elected representatives is loath to cross the regulatory line and impose *any* restriction on the treatment of animals *during* the experiment itself.

Fortunately, the US system of government has been set up to give animal protectionists another legislative option. Our government gives very specific powers to our national government and leaves all other powers to the state. Historically, the federal government has seized upon and expanded its zone of jurisdiction. This expansion has allowed the federal government to regulate the area of animal experimentation under its Commerce Clause powers of the US constitution. It regulates the activities under the jurisdiction of the National Institutes of Health because of the funding given to individual research institutions. However, this does not limit the power that the individual states have for regulating activity within their jurisdictional foundation.

*Holly Hazard is an attorney specializing in Animal Rights Law at Galvin, Stanley of Washington, D.C.

Because of our frustration with the federal system, many animal protectionists have turned to state legislatures as a forum for increased attention and protection for laboratory animals. Although a coordinated effort throughout our state system of government is relatively new, these efforts have given the animal rights community a glimpse of success.

In 1989, the political reception for increased protection for animals in laboratories is shifting from blind acquiescence on the part of legislators to at least a critical perspective on the evidence of waste and abuse in our laboratories. These changes are taking place at both the federal and state levels, but are more pronounced in state initiatives. The power structure at the federal level presents a more difficult challenge, particularly as we continue to press for reform *during* experimentation as well as before and after.

FEDERAL DEVELOPMENTS

The federal system of govenment in the USA has shown itself to be particularly ineffectual in addressing the social concerns of our society. Our federal system is set up to give control of issues to one or two Congressional Representatives. Many of these Representatives owe their continued presence in Congress to the participation of special interests in their campaigns. These interests then gain tremendous access to these few Representatives who have complete freedom to decide if a bill is given a hearing or is voted on by a committee.

Nowhere is this influence more apparent than it is on health research issues. Political action committees (PACs) representing the health industries contributed over $27 million to the election campaigns of Congressional House and Senate members in the last five years.[1] These same members sit on the committees making decisions regarding funding and regulation of our nation's laboratories.

The animal rights community has had difficulty in getting a hearing before these committees, much less in having committees vote its bills out to the floor. This problem has been most apparent in the House Sub-Committee on the Health and the Environment. This committee controls all the issues related to the National Institutes of Health and most health regulatory agencies. The committee has been extremely unsympathetic to our concerns. In the last session of Congress, only the Consumer Products Safe Testing Act was given a hearing in this committee and this hearing took place so late in the session that the bill was doomed to failure before the testimony was heard.

The bill, which would restrict government regulation of product testing, has been reintroduced in the 101st Congress. The bill would require that federal agencies ignore the results of the classical LD50 test and review all regulations and guidelines that require or recommend any acute toxicity test. Unless continued animal experiments can be justified due to the lack of alternatives, regulatory agencies would be required to change their regulations to allow non-animal alternatives to the acute toxicity animal studies.

Other bills reintroduced in the 101st Congress include the Information, Dissemination and Research Accountability Act. This bill simply calls for the dissemination of biomedical information through modern methods of science and technology to prevent the duplication of experiments on live animals and for other purposes. In a world of finite resources, this bill would seem to be an appropriate compromise between the interests of the animal rights community and the research community.

The National Institutes of Health, however, instigated a campaign in the last session to inflate grossly the cost of this legislation and to call for opposition to the bill in the name of fiscal conservativeness. The National Institutes of Health stated that the implementation of this legislation could cost the federal government as much as $2 billion to $6 billion. The bill has now been amended to exclude the translation of foreign journals. The Congressional Budget Office recently estimated the cost of the legislation to be approximately $38 million. This figure does not take into account the amount of money that will be saved by the government by rejecting grant applications for research found to be duplicative.

Congress also addressed the use of pound animals in biomedical research in the session of Congress ending in December 1988. As noted in previous chapters, the USA is one of the few remaining nations to allow pound or shelter animals to be used by the biomedical research industry. This bill was the focus of a determined campaign by the National Association for Biomedical Research. This organization is the leading lobby for continued animal research. Although this Association was a frequent participant in negotiations on both the House and Senate side to work out some compromise on the legislation, it was also actively involved in a public relations campaign to distort the intentions of the bill and to rally public support around the question of animal research in general, ignoring the statistical evidence that pound animals are used in a small minority of research protocols.

The Senate sponsor of the bill, realizing that his original legislation was doomed to fail, introduced legislation that would simply prohibit dealers from acquiring dogs from auction sales. This obviously would do nothing to help pound or shelter animals but would have been successful in stemming the theft of dogs or cats for auction destined to be sold later as pound animals to research facilities. When this legislation was brought to the House for hearings, the animal rights community presented a fractured picture of support and the medical research industry was able to defeat this bill. No similar legislation has been introduced in the 101st Congress.

The 101st Congress continues to grapple with issues related to the Animal Welfare Act as the Agriculture Department comes before the Congress for appropriations for enforcement of the Act. Until recently, the Department had asked for no funding to enforce the Act. Of late, the Department has become somewhat more interested in undertaking a responsible approach to its duties under the Act. As of Spring 1989, it has still not finalized the 1985 Amendments to the Act strengthening the protections given to laboratory animals.[2] Litigation is pending to force the Department to implement these changes.

Because of the growing frustration of the animal rights community with the lack of interest on the part of the Department of Agriculture or the research establishment to ensure that the modest provisions of the Act are upheld, Congressman Rose introduced what became known as 'The Standing Bill' in the 100th Congress. This bill would have allowed interested individuals and humane societies to sue the Department of Agriculture for non-enforcement of the Animal Welfare Act. The impetus for the bill was the growing realization by the animal rights community that the overwhelming majority of cases involving serious deficiencies of the Act were uncovered by individuals in the animal protection movement and not by the Department of Agriculture. At hearings on the Standing Bill in the Judiciary Committee in the House, Representative Barney Frank scolded the Department of Agriculture and medical societies for their lack of initiative in this area and told the opponents of the bill that they should give some 'hard thought' to what might be done to improve enforcement in this area.[3] This bill has not yet been reintroduced in the 101st Congress.

The Consumer Products Safe Testing Act, the Standing Bill and other legislative initiatives do not represent any revolutionary change in animal protection. Yet, the Congress has been sluggish in giving these bills a forum despite unprecedented constituent mail on the issues. As this lack of interest has become more apparent, animal protection groups have turned to other forums. Some have gone to the press and others have turned to consumer pressure. Many, however, have turned to state legislatures as a voice for their concerns.

STATE LEGISLATION

Because our system of government gives specific powers to the federal government and all remaining powers to the states, and because the states have historical jurisdiction over the humane treatment of animals, this forum is ripe for legislative initiatives to increase the protection of animals.

One obvious avenue for challenging research practices using state law is through the state anti-cruelty statutes. Every state has an anti-cruelty law. Although some of these laws specifically exempt research institutions from the provisions of the anti-cruelty statutes, many do not. A few states, such as Maryland, explicitly include research institutions.[4]

Interestingly, the inclusion by the people of Maryland of research institutions in the state anti-cruelty statute came about as a result of the first prosecution of a researcher for animal cruelty in the USA. In 1981, Edward Taub of the Institute for Behavioral Research was prosecuted for violations of the state anti-cruelty act after his research practices were uncovered by a student hired as a research assistant in his laboratory. This case was reversed at the appellate level.[5] The appellate court explained that the legislature had not intended to include federal research programs under the state anti-cruelty code. The legislature quickly moved to make its intentions clear and amended the anti-cruelty statute to include explicitly laboratories under its provisions.

Of course, it was too late to protect the animals in Dr Taub's laboratory. This statute, however, and others like it, should serve as a deterrent to researchers in other states. Currently, research is exempt from prosecution for cruelty under the state anti-cruelty laws in 22 states. Animal research is not mentioned (and therefore open to a challenge) in 25 states.

Another state-imposed restriction on research is in the growing number of states challenging the use of pound animals in medical experiments. As mentioned under 'Federal Developments', animal protectionists have been working on this legislation at the federal level. Efforts at the state level, however, have met with some success.

State activists have challenged the use of pound animals in medical experiments on several grounds. They have argued that the use of 'random source' animals creates flawed research because of the variables attendant in selecting an animal of unknown origin. A more compelling argument, and one that has been used successfully in several state legislatures, is that our local pounds and shelters should not be used as warehouses for the medical research industry. Legislators have agreed that individuals, many of whom donate their time and money to protect animals from neglect and abuse, should not be put in the position of being unwilling accomplices in the procurement of animals for medical research.

Another successful argument has been that individuals are less likely to want to turn in stray animals if they know that there is a chance that these animals will be turned over to research facilities. This threat discourages people from turning over neglected animals to humane societies and shelters.

A growing number of states have agreed with animal protectionists that some restrictions on the use of shelter animals in research are appropriate. Currently, 13 states prohibit the release of animals from pounds for the purposes of medical research. Massachusetts has the strongest statute prohibiting the release of animals. This legislation prohibits the use of pound animals from *any* source so that researchers may not cross state lines and obtain animals from the surrounding states.

A few states actually require the release for research. This puts humanitarians in an untenable position. They must choose between ignoring the problems associated with strays and other neglected animals or organizing a shelter and being subject to the collection of animals by medical research institutions.

The majority of states, however, have no mention of the release of pound animals in their statute. This leaves the decision for release up to individual shelter managers. Some shelters have actually lobbied against the animal protectionists trying to prohibit release for research use because of the profits received by the shelter for the animals surrendered to research institutions.

One of the most exciting developments in legislative advocacy for increased animal protection in the last few years has been the virtual explosion of state initiatives to restrict the use of animals in product testing experiments. These legislative initiatives represent the first large-scale effort to impose restrictions on the kinds of experiments that society should condone.

Maryland led the way by introducing a bill that would ban the Draize rabbit eye irritancy test within the state. In the first session that this bill was introduced, it sailed through the Senate and was successfully voted out of the House, only to be killed in a conference committee. In the next two sessions it had a much more difficult time. The Governor of Maryland has now set up a task force to explore this issue in depth and to make recommendations back to the government by the summer of 1989.

- Legislation to ban the Draize and/or LD50 test is currently pending in at least five other states. In New Jersey, a bill to ban the Draize test has passed the Senate and has been voted favorably out of the Committee in the House. As of May, 1989 it is pending on the floor of the House. In April, 1989 the Illinois legislature quickly passed a bill out of the Judiciary Committees in both the House and Senate which would ban the Draize eye irritancy or skin irritancy test. In the

1989–1990 session, the California legislature has two bills pending related to this issue. One bill calls for the labelling of products that are tested on animals to enable the consumer to make a choice for a more compassionate lifestyle as he or she chooses. The bill calls for certain products to be labelled accordingly if they have been tested on animals. Another California bill would ban the LD50 and Draize test.

The state legislatures present an excellent opportunity for change in the area of laboratory animal experimentation because the health lobbies are not as organized and have not been as successful in ingratiating themselves with the leadership of the legislatures. In fact, in hearing after hearing many legislators have been hostile to the positions taken by medical researchers and product testing representatives in response to proposals to restrict animal testing. Representatives of industry and research institutions have been called arrogant and unprepared as they attempt to present their positions on these issues.

An obvious course of action for the medical and product industries is to become as sophisticated at the state level as they have been in Congress. I have no doubt that this will occur in the not too distant future. Until then, however, animal protectionists have an excellent opportunity to have an equal voice in the legislative process, at least at one level of government. This presents a challenge that animal protectionists should not ignore and, as the record indicates, have in fact taken great advantage of.

CONCLUSION

Our federal system of government is unlikely to change significantly in the near future. The Committee system and PAC financing will continue to stifle change on all social justice issues. What is encouraging about the legislative process, however, is that the animal protection community is becoming more sophisticated in the process and more respected by the legislators. No mystery exists as to why the research industry has been as successful as it has with legislation in the past. As the animal protection community develops its skills as a lobbyist and a political entity, it will become more influential and successful in the federal process.

At the state level we have already seen some glimpses of success. If we continue to present reasonable and fair options to our legislators this forum should also present us with some success in our efforts to protect animals in laboratories.

NOTES AND REFERENCES

1. Anon., How health PAC's spend millions to influence elections, *The Washington Post*, 21 March 1989, Health Section, at 12.
2. 7 USC 2131, *et seq.*
3. *Hearing Before the SubCommittee on Administrative Law and Governmental Relations of the Committee on the Judiciary, House of Representatives, One Hundreth Congress, on H.R. 1770, Animal Welfare Act, September 16, 1988*, Serial No. 76.
4. Ann. Code of Md. Art, 27 §59.
5. *State v. Taub*, 296 Md. 439, 463 A.2d 819 (1983).

Index

261